CREATE YOUR UNIVERSAL LEGACY

DAVID P. SARMIENTO

Cover image courtesy of NASA.

ISBNs: 978-1-7364531-0-0 (paperback); 978-1-7364531-1-7 (casebound); 978-1-7364531-2-4 (e-book)

I dedicate this written work to those who desire to make a change in this world by improving the general welfare all humanity needs. I thank Jesus Christ for the many opportunities to travel internationally gaining experiences and visiting religious sites in the Holy Land of Israel. I hope this book will enable me to aid the Monastery of the Cross and the Holy Order of the Cross seeking to serve humanity. May the Lord's will be done on Earth as it has been done in heaven. Amen!

CONTENTS

FOREWORD

This literature is for humanity's open participation to correct our environmental failures and to improve our social relationships for the advancement of the human race.

PREFACE

This book's contents are intended to inspire readers to go forth with a mindset that they can change the universe through their implementation of hard work. Everyone chooses their own path for transformation and success. Your sparks of actions or inactions will be the catalyst that inspires others around you to begin a new life. As you change humanity for a positive future don't forget that living creatures in their natural environment also need your compassionate caring.

1

UNITED SECURITY ACT (USA)

Mandatory military training for all US civilians would increase domestic and international defenses opposing terrorism. The US government must improve American national security by mandatorily training all citizens, preparing them domestically and internationally for potential dangers. Terrorist groups are willing to use conventional, nuclear, chemical, autonomous, cyber, and biological weapon systems against all American allegiances. The National Emergency Security Task Force (NEST) adds two layers of additional security over potential threats. NEST minimizes the dangers from conventional, nuclear, chemical, cyber, and biological autonomous warfare systems capable of plaguing Americans.

Terrorists have attacked nightclubs, marathons, skyscrapers, conference facilities, schools, churches, hotels, military bases, and crowded streets, constantly changing their tactical strategies to inflict harm. With so many different forms of violence and moral values regrettably diminishing from society and disrupting peace, our governments must change to bring about peace. Governments are people, so America's social and financial problems reside upon the shoulders of poor leadership losing control of their integrity. American citizens are opposing the faces, hearts, and minds of government employees because there's been no compassion from elected officials for those most overlooked. The American government has sought a position of superiority to be above protesting civilians, Civilians are only acknowledging governmental injustices which are being cover up by the decision makers who experience no suffering as they subject citizens to extremely harsh brutalities. America and the world can only change when those in power willingly decide to change along with everyone else making social changes for transformation.

A government of minds placing themselves above others on a constant basis is a superiority complex that resides in America. "All lives matter" should be the international policy to promote peace. Universally, there should be no more government structures designed to allow some people benefits and not allow others the same benefits. This aforementioned warring strategy has been used as a hate tactic for way too long to oppress all ethnic minorities. No nation can endure refusing to listen to truth. For all nations must obey righteous messages of peace that address its social and economic problems.

American civilians must be trained to aid police, fire departments, hospitals, national guard units, national agriculture, and many forms of infrastructure before catastrophic attacks occur. Granting military training to all Americans unites the nation to be invested in its country instead of being divided against its leaders. American leaders cannot continue their one-way ideology, wanting everything for themselves and those whom they permit to be privileged. Accepting citizens—even felons—and giving them a second chance with more opportunities having no restrictions strengthens a nation. Unlimited opportunities for citizens allows a nation to move in the right direction.

The Holy Bible says to forgive seventy times seven for offenses. America's national system has adopted a policy of no forgiveness with the background check system enhancing sins into a realm of never being forgiven. Federal background check systems grant—double jeopardy on all accounts forgiving no one. Felons, under the age of eighteen, raised in communities where crime and dysfunction exist, shouldn't have their entire lives destroyed by federal background check systems.

What nation will be greater: one that forgives (100) one hundred percent or one that holds people's offenses against them for the entire duration of their lives? I'd like my nation to be the greatest nation on Earth. One hundred (100) percent forgiveness will be where my nation supersedes any other nation. My nation will be more loved, respected, and held in greater esteem than any other nation on Earth because its people are completely forgiven for their offenses. Citizens will honor and support the eternal history of my nation above any other nation because they can prosper without repetitive denials.

Eternally, people will support the history of my nation while other nations' histories collapse destructively being a testament to their lack of forgiveness. Billions of people desire fruitful leadership that understands their needs, desires, and aspirations. Governments who cannot relate to their citizens' needs will quickly find themselves rejected, and adding to their eternal legacy fading into oblivion.

Political groups collectively have clout to manipulate laws in their favor like Greek society alumni and private organizations scratching each other's asses. The electoral college parties who support laws that knowingly do not benefit the majority of citizens remains a national problem. The corruption of electoral college leadership running our nation's political hierarchy has increased America's national debt into the trillions of dollars unable to balance budgets with they're combined intelligence. Americans should never admire or hold in high esteem the ignorance of government leaders whose decisions decisively hurt humanity year after year.

National debt is something that shouldn't be tolerated for when wars and unprecedented events arise, these catastrophes can quickly cost trillions of dollars. If America is ever be involved in another world war, civilians previously trained with one or two years of compulsory military service will find themselves better prepared to enter the draft.

Rapidly training Americans and then rushing them off into foreign guerrilla warfare conflicts unprepared was not the best strategy in America's past. American casualties were higher in wars with drafts than in any other war utilizing trained soldiers. American citizens are not protected from other citizens, and all citizens have no protection from terrorist attacks. Terrorists are successful in their undertakings as society attests to the security failures in American history. Providing military training before wars begin can quickly determine where citizens stand morally, ethically, spiritually, psychologically, physically, and educationally for leadership positions.

Not all officers are good leaders, and some officers shouldn't be in positions of leadership because of their inability to deal with others properly. Sorting out who should lead from who shouldn't lead is very important in any nation that seeks to prevent criminal prosecutions and abuse problems beforehand. Terrorism cannot be conquered if all Americans are not united. Uniting Americans means making equal opportunities for all Americans. We must eradicate all denials originating from national programs such as military branches, government offices, federal agencies, and departments of licensing, whether state or federal. Many Americans believe racism exists; changing policies makes things right.

The equality of being trained in programs like the National Emergency Security Task Force instills positive values in recruited civilians in all fifty states. The National Emergency Strategic Task Force will be positioned to aid in disasters separate from existing National Guard troops. One example is when recent hurricanes caused major flooding, groups separate from the government stood up and aided civilians more quickly than trained military units. NEST shall operate independently from any other form of government in emergencies even in full-scale war, natural disasters, and massive terror attacks.

The core value of the National Emergency Strategic Task Force is aiding those in need. In extreme situations, NEST can evoke powers from its command structure to perform detainments, arrests, and imprisonments to maintain peace. The National Emergency Security Task Force shall operate much like a militia with its own judicial system. As America spies on everyone, it seems like everyone is spying on Americans. Many nations viewing classified or declassified American files from hackers no longer trust the United States and are committed to Cold War mentalities indefinitely. The Department of Homeland Security is practicing espionage like the Central Intelligence Agency, and the National Security Agency. All this spying only means agencies have opposition from foreign powers and jihadists willing to establish their own spy agencies.

In this world where everyone is watching everyone, the only alternative is the National Emergency Security Task Force. The National Emergency Security Task Force would autonomously operate separately from the three American government powers doing poor jobs at national security. The government did not protect the United States from the September 11th, 2001, terrorist attacks, in addition to so many other terrorist attacks since then. The National Emergency Strategic Task Force would operate in any areas where government training or systems have failed. Since the objective of the National Emergency Strategic Task Force is to protect all forms of life, any groups opposing it would be inherently endangering lives. During emergencies the American government inadvertently unable to fulfill its duty protecting citizens would have NEST positioned for making righteous judgments for the benefit of the nation.

NEST will uphold honorable values, moral rules of conduct, and righteous decisions of integrity having higher standards than any organization. The Constitution of the United States was established with the sole purpose of empowering its citizens against injustices. The American government has been enforcing a threatening behavior and contributing to injustices for a very long time, by not obeying the US Constitution.

The federal government has strived to strip away power from Americans, so that the US Constitution is being undermined to disenfranchise people through hateful force and the manipulating of laws to oppress civilians. Force enacted upon American citizens by the federal government through many unjust tactics warrants prosecution by the department of justice for offenses done in criminality to sabotage freedom. Millions of Americans and billions of people worldwide already know that the justice department is compromised and has been for a very long time. Exclusionary methods practiced by the US government to prevent citizens from obtaining full rights while enabling these privileges to be hoarded as benefits for certain government employees is oppression. These oppressions are aimed at harming American citizens, by disenfranchising citizens financially with pay reductions and exclusions. American citizens are experiencing hateful inequality, bias separation, permit denials, preventative employment, property extortion, license banning, and restrictive educational advancements beneath the United States' government's oppression. Some Americans believe the FBI is one o the worst hate groups having white supremacists from the 1900s passing the torch of camouflaged racism along from one generation to the next. How is it that so many forms of labeling of people badly can be utilized by it's members dispersing denials as a means of oppressing minorities? The Civil rights movement was only in 1954. Ask yourself this question can a racist person act like your friend and hide the fact they're planning to assault you surreptitiously at the same time? So many unsolved crimes exist and I believe this can only be possible because on the inside within government including the FBI there are criminals hiding in uniform. There are racist groups hiding within the American government wearing suits, badges, and military uniforms without any intention of ever leaving their powerful well paid positions.

Government employees using strategies to funnel wealth and resources into their control while excluding citizens from resources is one form of American hatred practiced. This aforementioned method of hate is used to keep taxpayers from being able to prosper while enabling those practicing racism to reap greater financial gains than minorities. Exclusionary hatred denying citizens what they seek is a massive problem in America because denials lead to citizens' dreams being destroyed by their own governmental leaders. Who wants to support a government body that destroys its citizens' dreams and their children's dreams? The US government's employees have become experts in destroying the dreams of American citizens through denials.

Since destroying the dreams of others has become the legacy America has established for itself with millions of minorities disenfranchised, and it's time to make changes. When one racial group is supported more than any other ethnicity, this creates resentment as well as hatred. We see hatred brewing in American schools with students murdering each other and the parents of both groups being at odds with each other. Voicing these concerns in documentation hasn't corrected the unrighteousness found in American leaders harming the lives of other ethnicities. Preventing these merciless small-scale wars is important to those of us seeking peace by making known the concerns so all the issues can be corrected. The problem is that hateful government leaders refuse to correct the problems voiced by neglected citizens being victimized. The American government's failure to disperse justice is why the hatred builds to the rupturing point creating the murders in our society.

Governments that practice hatred will eventually reap their own internal collapse. The creation of newer, better government systems will replace past eras using outdated political methods. Modern global changes require more oversight to ensure problems have advancements made frequently on issues needing correction. Biblically speaking Americans targeting wealthy families and military families could be the future of America if changes don't take place for the future of America. University and high school murders show that hatred is rampant in America already. Yes, these violent murderous occurrences in education systems point to huge problems of mistreatment contributing to aggression and mental illness. America has millions of people with mental illness inwardly concealing their hatred. What will be the result when terrorists recruit mentally ill Americans for violent attacks on military families or government as a possible scenario? My point is, I'm witnessing so many problems arise that if no changes are made, small problems can quickly become humongous problems. America is losing its footing domestically as well as internationally, by being less reliable and less admired in the eyes of many. Illegal drug use in America isn't helping the mental illness problem either.

America's employees working in government have been participating in financial warfare tactics by constantly spying for financial gain and stealing people's livelihood. Some American agencies have conducted imprisonments to silence intelligent citizens voices to prevent new leaders from emerging by practicing hatred. Overlooking these hate crimes is a failure of the US Department of Justice, which is ignoring citizens' rights by siding with government employees possessing criminal mentalities to accomplish unjust offenses against minority citizens. Nationwide, citizens are offended by government officers collaboratively using tyrannical methods to plan offenses contrary to justice. Government officers undermining the thoughts, desires, requests, and complaints of citizens has concretely proven the biased discourse of members holding government positions. Government employees' techniques to blatantly disperse denials is a hate tactics being practiced to oppress citizens with criminal behavior. Not enough veterans, soldiers, officers, or government employees are being sent to prison with long term incarceration for crimes against citizens.

The American government's employees practicing a tactical mechanism of robbing citizens of their time, resources, health, and wealthy information in every tangent of advancement through denials is hatred. The American government's use of diversionary tactics to prevent achievements is a hate method practiced more often on minorities than on any other ethnicity. These aforementioned criminalistic evils of discriminatory procedures, strategies, and formulated approaches of prevention are often used by America's government and military officers to suppress civilians.

Government employees conducting evil schemes to rob citizens of licenses, rights, and advancement in their lives is directly opposite to constitutional law: "with liberty and justice for all." America has become more divided every day; with its leaders refusing to grant citizens and felons new lives, which are open to all opportunities. How can any government hoard privileges and deny others the same privileges or opportunities which would be used to improve lives? Many American's have begun to hate the faces of the people running the government in the United States because their lack of empathy, injustice, and corruption.

The American government's happiness has been to deny its citizens equal opportunities and to cover it up preventing dreams from becoming a reality which is evil. Americans don't desire to see government soldiers adorned with: merits, benefits, higher pay, and unrestricted opportunities while civilians have been rejected from any participation. To wave an American flag or a United States Constitution in the faces of millions of people offering rights, then to deny those same rights strategically through employees is hypocrisy. I'm pointing out these errors throughout the foundations of America, where falsehood dwells in the hearts and minds of corrupt people with powerful occupations intentionally harming others.

I can find millions of Hispanics, Asians, Indians, Hawaiians, Pacific Islanders, Africans, Native Americans, and those of other ethnicities who agree with my words about the insincerity, dishonesty, dissimulation, and mendacity in the constitutional dispersion of rights. To improve American policy, leaders must begin with all federal departments and agencies changing their decision-making to include yes responses to requests from citizens. Not making changes now will have powerful repercussions as more people start looking for alternative rule to American governmental positions on a global scale. No government these days can suppress global communications between ethnicities and cultures. America can attempt to use technology to control global communications restricting voices from being heard to war against ethnicities. The fact is people from many ethnicities already know how the hearts of this nation's discriminatory people are in ruling structures of America's government possessing a ill will toward everyone else unprivileged.

Any government that wants to suppress or control global communications is a dictatorship. Hopefully the American government hasn't transformed itself into becoming the corruption it once opposed. Training an entire nation unites all civilians, increasing the awareness that they are the potential targets of terrorists. Terrorists cunningly train their adolescent offspring in warfare tactics, and techniques.

Americans must stay alert to countries that have mandatory military conscription training to not fall behind. While other nation's have a willingness to strengthen they're people to have brigades, divisions, fleets, and air forces with military training the American government continues to weaken it's civilian population with denials for military training. What does it say about American leadership if the enemies of America train their own, and America's leaders haven't trained our citizens who are attacked by our enemies? The fact is Americans civilians have died from terrorist attacks even after 911, while trained soldiers remain safe and secure in fortified military bases.

I feel saddened for the many generations of people in America who've had their dreams destroyed by employees of the American government answering no to requests while refusing to uplift civilians. American soldiers, pilots, sailors, and personnel have traveled to foreign nations to train with foreigners who are better educated and better trained than America's citizens. America's citizens receive less from their government leaders, than foreign nations who give they're people military training, practical survival experience, and continued educational opportunities in conscription. Refusing to educate American civilians who possess fewer occupational opportunities than foreigners, doesn't support America's position of being the best nation to suffering Americans denied advancement. Denying life, liberty, and happiness to civilians through government offices conducting a preventive acceptance has stopped the occupational attainment of citizens; and is active criminality taking place against American citizens being hated. Our government's singular most professional skill seems to be creating experts in being dick heads all the time. I don"t know how someone could enjoy being a dick all the time but some people like having their picture drawn this way for they're professional portfolio of life accomplishments (:)).

The union of elected government officials preventing Americans from benefiting from the US Constitution and the Bill of Rights proves that discrimination is being used to halt the attainment of equality. America's elected officials are warring financially, socially, and educationally against the mass majority of American citizens in this nation.

Nations practicing communism, totalitarianism, socialism, monarchies, dictatorships, anarchism, tribalism, despotism, and feudalism are giving their citizens military training. America's government denying its citizens military training and educational opportunities has absolutely proven that America is divided at its core. American government leaders should embrace and advance their citizens, not make them less educated than foreigners who are granting their own people the training they desire. America's leadership has failed past as well as current generations because of the discrimination dwelling in the hearts of people within this government and the ideology "for us and not for you."

Denying citizens happiness will lead to war unless changes are made in the US government. America is not a monarchy, so lording over the lives of citizens isn't behavior police, military personnel, or government leaders should practice three hundred sixty-five days a year. American soldiers, police officers, and government employees shouldn't transgress on a daily basis, wanting to be above others; for that's one problem adding to social conflicts. American citizens can refuse to give love, relationships, and socialization to all soldiers or government employees denying citizens' dreams. Making government employees to be outcasts to be hated if no changes are made could be common place in mainstream society as injustices continue. If positive changes aren't being made to give civilians more prosperity there could be a more massive American movement. I say these things because the Holy Bible foretells that the love of many will wax cold. Today's teenagers and adults are killing one another so this is only the beginning; for it shall get much worse if leaders refuse to heed commands to change. Millions of women in America and around the world refusing to give birth to the seed of corrupt American politicians, soldiers, officers, and government employees could be a movement. Eliminating future generations of evil people maybe exactly what society needs to balance this world's problems. Yahweh sent a Great Biblical flood to rid the world of corruption maybe it needs to happen again but in an entirely different way.

"What goes around comes around" is the old saying, so the US government cannot continue to practice hatred, distributing denials. American policies must be changed in order to form a more perfect union. American posterity must allow everyone to obtain equal opportunities, not just some who, through multiple generations, have empowered themselves to oppress the weaker citizens in America. Americans do not want any more school shootings, bombings, protests, police murders, and unrest. Americans do not want Americans killing one another in grand-scale civil war conflicts.

Increasing the strength of some and not all is a very cruel way to punish the weak making citizens vulnerable during civil unrest as well as during a pandemic. America's government has lost righteous virtue in strengthening the weak. The Lord Jesus Christ healed the weak, and aided the poor feeding the hungry who were destitute. Americans have grown arrogant, offending the most high in the ecclesiastical realms by endorsing those who shouldn't be endorsed. The American government's resource preparations have increased the strength of soldiers but not all citizens, which is discrimination. Those who do not believe training all Americans is the answer are the individuals unfit to be called leaders. Corrupt leaders and government employees have the intent to weaken this nation by harming its citizens through a depriving spirit of hatred to disallow civilians to have. The American government doesn't want citizens protecting ourselves, and they refuse for civilians to be given equality to have possessions because all their angles are designed to disenfranchise us from prosperity. Weakening citizens is not honorable but this is what the American government enables by its endorsement of devious people with ill will towards others. Americans are not naive; citizens for we have the answers to guide our children down a more productive path to adulthood. Terrorism has become a part of American lives, along with so many incidents of civilians being attacked. Countering terrorism requires training all civilians to oppose extremist alliances hiding within, by providing educational experiences that civilians can benefit from to protect themselves from dangers. Not uniting all Americans gives terrorists a foothold to divide citizens for violent paths to attack their fellow citizens. Terrorists will never stop targeting soldiers, police officers, and government leaders' including their families.

American leaders must recognize the truth that civilians are potential targets needing military training. Domestic terrorism and international terrorism include the mass shootings or bombings in which civilians remain the victims who are most vulnerable. American training can also provide opportunities for conscientious objectors to serve in alternative forms of noncombatant training. Finding out now, instead of later, who the conscientious objectors are is wise, so when a conflict arises, leaders need not be distracted with these issues in time of service. Why have so many American citizens opposed the police, state governments, and federal agencies? Maybe the answer is that citizens are feeling they're being excluded from crucial training and educational opportunities by corrupt leaders who place themselves oppressively above everyone else. Not wanting civilians to have any experiences and aiding those who commit crime against civilians has been an angle the United States government has covered up. Citizens only have so much of a tolerance level especially when their financial prosperity is being destroyed and they're being murdered.

America hasn't been teaching honorable virtues, responsibility, or, yes, advanced educational studies to many of its delinquent citizens or adults who need boot camp instructions. Boot camps early on can be precursors to serving in other capacities, and we shouldn't quickly dismiss behavioral changes that are part of character growth to becoming better adults. Parents that never taught their children right from wrong will have children that need more guidance and correction. Children who had parents voicing truth constantly throughout their child's upbringing into adolescence will require less correction and more focus oriented guidance.

Toxic veterans need to be removed from the equation of teaching our youths for they're terrible teachers. Toxic veterans think they're therapists or doctors while concealing addictions as they attempt to brainwash others they think they're providing treatment to. The LGBTQIA community having addictions and thinking they're medical professionals is a very dangerous situation for children or adults in out communities.

The LGBTQIA community getting access physically, and socially to children or teenagers is dangerous especially because they steal personal belonging to get into the minds of people they want to treat. The LGBTQIA community of addicts believe they're authorized to treat others and has led to many children being molested as well as adults being harmed. The United States military has given psychology training to delusional soldiers who are mentally disturbed individuals possessing distorted irrational thoughts. Some of these soldiers leave military service or are discharged and arrive back into our communities with a veterans administration identification. The real danger is when members of our community and government help to finance or facilitate these individuals living surreptitious double lives. Veterans with mental problems have gotten community therapists or doctors they've seen in on the racket to examine the lives of others not needing counseling. A game for delusional people with mental illness is to take the focus off themselves and cast attention upon others as being mentally ill. Sneaking behind a victim's back and stealing to tamper with their lives creates chaos that the mentally ill person capitalizes off of to disrupt lives toxically. A narcissist lives with a personality to always hide what they're doing as they harm the lives of others as a covert operation. How some narcissists think; "because I was in the military I can do whatever the hell I want, even if my actions sabotage your life, and disenfranchise you from any knowledge or wealth your hands created."

The misguiding and manipulating of a professional's mind who analyzes others, will be met by the formulated counter offensive thats well thought up by the narcissist. The narcissist cognitively attempts to outthink professionals as part of their ongoing delusional behavior wrapped up in their superiority complex to supersede others. When sober non drug using professionals give correction it's good, but possessing a degree of restraint is also part of becoming a respected leader showing empathy. Some military officers want to be honored while they've lived dishonorably in the eyes of many. Many officers, federal agents, and elected officials have become exposure for the international news for the corruption they've been involved in during criminal cases in their community. We must scrutinize and reevaluate leaders who've endorsed bad veterans creating chaos in our communities. Sexual scandals seem to have a consistency in American news. American scandals in the international news have alerted psychoanalysts from many countries to analyze America's leadership behaviors associated with mental illness that have been displayed.

American leaders who aren't educated to possess an understanding of religious laws and ethical codes such as chivalry, courtly manners, and the honorable nobility shouldn't be in elected offices. Elected offices warrant the highest standards of morality to be viewed by the global community. America's leadership shouldn't disgrace the international community with conflict but rather show that our behaviors uphold the highest level of professionalism to not be offensive while remaining at peace. Children worldwide are growing up seeing elected leaders behave without morals. Its not good for parents trying to teach their children right from wrong to see national leaders sinning while the world witnesses bureaucratic disorganization. How can Americans claim to be better than any other nation's people or infrastructure? America's leadership is sinning in the international news not living a hundred percent devoted to their wives or country? How many billions of people are influenced wrongly by the actions or inactions of Americans? We need to do a better job teaching, because the way military academies teach is the way we'll correct this nation's problems. Students need to get back on track and they need to start practicing morality.

Modern times require mandatory military training for a minimum of one year for all civilians. Many countries around the world make military training mandatory. Remember that those partaking in terrorist activities and organized crime are teaching their offspring since adolescence to make war against laws from an early age. Gangs networks aren't going to stop criminal activities or narcotic distribution till we stop them with massive incarceration. The mistake of our nation's past conflicts was rushing young men to train quickly and then hurrying them off to war. These historical American deaths were very costly caused by unworthy officers in leadership positions who refused to acknowledge they shouldn't have given orders for young soldiers to rush off into battle ill prepared. Military officers unwilling to admit there's someone better to lead is a big problem in our United States military branches. Warfare textbooks haven't taught everything needed to run a theatrical military campaign for there's just too many commanders lacking common sense because they're followers and not leaders. Officers frequently transgressing against their superiors is a problem, while they think they should lead, but are inadvertently harming others because they're mental illness. Military officers disguising their mental illness is a problem in the United States. Medical personnel that have hidden their mental conditions which require a psychiatric hospitalization haven't received it outside of military service. We've witnessed military veterans participating in mass shootings, euthanasias, and the harming of patients physically or psychologically while being improperly credentialed. Nonmilitary professional psychologists and psychiatrists need to evaluate soldiers to determine if they're fit to reenter society, and not another military ranked professional soldier. Soldiers will always side with other soldiers so this thwarts the effectiveness of psychological or psychiatric review by licensed professionals. No veterans who are psychologists or psychiatrists should be analyzing soldiers to prevent the evaluations from being compromised by other factors outside of medicine.

Terrorists are not going to stop attacking civilians while American soldiers remain protected in fortified military bases. Americans must improve national security through national service rather than ignoring or doubting that weapons of mass destruction will be successfully used in the future. Terrorists have determined that World War III will be on American soil. The targeting of powerful families who will be holding their battleground is expected by terrorists. The attack on the Pentagon was just a precursor to future attacks planned against Americans on a massive scale. Increasing national security requires changing our thinking as well as our infrastructure.

In World War III, expect no degree of mercy and only the greatest cruelty, more callous than any in the history of human life on Earth. Americans can no longer follow outdated ideologies practiced centuries ago for our modern-day ways of running a government. Governing in modern times requires special skills promoting peace to reduce conflict by helping allies improve their general welfare systems abroad. Sharing knowledge promotes global welfare and enhances love that American's need for transcendence to establish a posterity in this changing world.

This United Security Act prevents the errors of the past by training Americans for a period of one year as a mandatory enlistment. Military training teaches many skills in audition to: codes of conduct, honorable service, discipline, and different forms of respect missing in millions of American lives. A united nation trained together is stronger and more unified. A nation in which police are disrespected, targeted, and dishonored can be prevented. The evaluating of all Americans for psychological conditions can be a multistep process through mandatory military training. Isolating mental illnesses and addressing problems early on in military training will prevent future violent occurrences in our society disrupting peace.

Evaluation and tolerance shall be benchmarks in improving the mental health conditions of civilians in the United States through mandatory training. The final part of mandatory military training will be instruction in the use of force, including rules of engagement. Etiquette will also be taught, in military training so trainees can transition into a variety of jobs, progressively adding honor to this nation. International trade and diplomacy require moral etiquette for business negotiations among all countries interacting peaceably.

How can minority Americans support their government's position when it puts forth denials and rejections creating unhappiness. Minorities have endured living with reduced prosperity for way to long over multiple generations? Before certain races are even born, many opposing Americans have begun the preplanned denial process preemptively, warring against infants dreams to racially disadvantage them. A nation that shares with all its citizens wipes away fear of reprisals because its people are invested in one another without favoritism being given to some people. Open military participation is the way to unite Americans and not to divide them.

America needs an additional means of gathering accurate information, and not just some government agents feeding commanders false intelligence to retain their jobs. With security lacking in many departments and agencies, the sharing of information isn't always necessary when deeper investigations are taking place. The ideology of taking what belongs to others is an invasion of privacy enacted by the federal government through the Patriot Act which is unconstitutional. The US Patriot Act is supported by soldiers accustomed to being thieves. Since the US government is bullying citizens using soldiers or police as enforcers to strip away a citizens rights this can be considered an act of war upon Americans. Any government constantly stealing from its citizens is a crime, and those supporting these activities are dictators enforcing unrighteousness hatred upon citizens. Mentally ill veterans aren't brilliant they've been concealing their crimes of how they acquired such knowledge hiding behind credentials and uniforms of their military past to enable their crimes. The United States government has been enabling these crimes by refusing to prosecute veterans thus being an active accomplice in home invasions, espionage, and theft harming citizens lives. This aforementioned reason is why American security has lapses because criminals have knowledge they shouldn't have. This is why we must replace employees of the American government because they no longer have integrity or credibility. Don't ever believe someone who steals and then at a later times replaces what they stole having copies of data hidden elsewhere.

The thefts enacted by the United States government have breached the separation of powers facilitating soldiers and veterans to continue committing crimes which warrant imprisonment. Rather than providing justice often the American government will side with the criminals they support to further the victimization of citizens to suppress people with hatred. If your an American experiencing this hatred from your own government's strategies to undermine your soul hail the creator Yahweh to intervene. Watergate methodologies used every day are hateful oppression. The domestic tranquility of government needs justice. Many American soldiers have murdered citizens and have rarely been punished even when alerted before the tragedies. I warned the FBI and alerted the police to crimes committed against me but they did nothing except let the criminals be as free as they desired enjoying riches stolen from me. Why are Americans putting less confidence in the police I just explained one reason. Oppression through theft leads to strife in communities, and adds conflict between citizens to the point of violence. The way to end violence between soldiers and citizens is to make them equals so everyone receives the same opportunities. The American government practicing sedition against the people of a nation is also a criminal act so lets help unite all Americans with mandatory military service. I don't believe most people get enjoyment going out and harming a numerous citizens especially when they've been peaceful for decades having rejected such thoughts for a very long time. I believe people commit murder because they're unhappy and someone helped create this unhappiness in their lives. Discovering the people who make others unhappy needs to be a campaign to prevent mass shootings or violence period. Many felons have guns illegally and don't go out killing people just because they're mad. Yes there are some felons who do snap and kill others but the motive to do this often is triggered by a series of problematic issues. The problems in our society need to surface and this can only be done when people are forced to communicate. Psychopathy is part of military training that all soldiers need for good communication skills and expressing their feelings. Really has police saying someone cannot have a gun stopped a felon from getting a gun if they wanted one? Many will say yes to the aforementioned statement but there are many felons with the means to kill people and don't do it. Felons often are strong and can easily murder two or even four people without a firearm if they wanted too. So just saying someone is a felon doesn't make them a person who's going to go commit every crime listed on a police officer's coding index. The focus to stop violence is to stop prying in people's lives and stop being thieves wanting to judge others or disenfranchise victims. Many people have the means to kill and don't; but when their victimized again and again this causes them to have a motive to commit murder. So when murders happen also ask the question who was participating in the victimization of another person? Don't just say to yourself it was a felon committing a murder because many felons are tolerant people having the power to murder yet have refrained from using it. Who are any of you to judge saying that a felon deserves no power it maybe Yahweh who gave power to the one whom all of you judged unrighteously. Many people have become felons according to judges and later in life it was discovered they didn't commit the crime so who are you to say they shouldn't have power? Some men or women have power and don't deserve it while others have power yahweh gave them with others unrighteous saying they shouldn't have it. Don't be so quick to deny the crowns off the heads of those whom Yahweh has deemed worthy to wear a crown of life. Many sinners Yahweh has forgiven completely giving them power but because those still sinning in society cannot accept that felons are forgiven being blessed with knowledge greater than themselves their super-ego narcissism victimizes the Lord's peaceful servants. Many felons have become peaceful servants of the Lord but the enter a society of people willing to violate their space and pry into their lives criminally. Each felon is different and therefore shouldn't be labeled identically which is what the federal government has done refusing to erase felon from their record. This is one reason America shall never endure because where no forgiveness resides very nigh will be the Devil who strives to oppose the sacrifice of Jesus Christ to forgive sins. America will not endure because it's United States Constitution was written eliminating the truth that Jesus Christ died for the sins of humanity and all those who believe. A government designed around not forgiving and exalting itself higher than Jesus Christ is destined to fail succumbing to destruction. If you listen to truth then we can only prolong the inevitable. For all those in government thinking they're thoughts supersede mine you haven't discovered that Yahweh sides with my decisions favoring me above yours. Yahweh only sides with me because I speak the truth even though I'm not perfect in my life either.

We live in an age when civilian teenagers need training to become better adults, so let's teach them righteous virtues so they will hold true to these values when they become adults. Exclusion from military training and educational programs is an old way of thinking. A government creating separation spiritually as well as physically in society follows after the Devil who divided angels in heaven with sin. With America's war on terror affecting past, present, and future American lives it's now time for citizens to have a law passed requiring a minimum one year of compulsory military service. When soldiers complete one year of service, they can be given the option to participate for a second year of service.

The second year of military training will quickly separate those who enjoy the military from those who would like to do some other occupation. The second year of service will be more advanced than the one-year compulsory service meant to unify Americans on common ground socially and educationally. With all Americans getting stronger on a better foundation of unity, then America will become a healthier nation, mentally fit with reduced health problems. Mandatory military training will reduce American cardiovascular diseases. Improving the health of American's is a positive move for this nation to reduce obesity and promote better fitness.

America neglects the needs of millions of Americans, and then imprisons millions of Americans through the judicial system, harming numerous lives. How can any nation be respected with such discriminatory hatred? Excluding repentant sinners from any employment they desire will be measured in heaven and on Earth for those judging who will be barred from heavenly duties for being sinful. The ideology people have that they can judge others and be given greater authority in heaven is false for Yahweh gives authority to those obeying his truth. Yahweh doesn't give authority to those refusing to obey his commandments. Treating everyone equally without exclusion from military programs creates camaraderie in a nation, instead of generating much division. Covering up the equal protection of laws with orchestrated judgments inflicting harm on minority American lives isn't a strategy that can continue in this nation. Current transgressions have reached a point at which tolerance is no longer accepted, and protests emerge creating confrontations with police or government officials.

America must accept its allies who also have compulsory training; this nation must recognize the truth that the enemy considers civilians wartime targets. If all Americans are given skills equally, there are no grounds to create division in America but, rather, unification. Terrorist tactics are constantly evolving in newer, ever-changing strategies. It's time for America to awaken and be better prepared giving citizens military training to aid this nation in many ways before the next terror attack. In my youth, I learned to be prepared; we need our citizens trained so they can help nationally or internationally when activated. The goal is for citizens to aid the police, fire departments, hospitals, National Guard units, transportation, agriculture, and many forms of infrastructure during war. Citizens can also help during natural disasters, and terrorist incidents unifying them to be better Americans.

Military training doesn't limit its structure to combat for history as well as other classes are taught. All citizens in military training must read the entire United States Constitution as well as the laws which they must abide by. There are many values citizens can adopt from military training. Values taught in military instruction can be respect for others, continued education, a strong work ethic, correct communication skills, discipline, and how to follow a chain of command. Realistically, all American children know how to fire weapons from the stacks of video games they play, so let's go a step further. We need to teach safety with responsibility—something that video games do not teach—and we need to do it through organized military training. Citizens need to be physically uplifted which reduces health-care expenses and make themselves more productive.

Denying citizens over multiple generations the rights afforded them in the Fourteenth Amendment to the US Constitution is "abridging privileges." US law has been manipulated by American judges, juries, state offices, and federal offices denying equal rights to citizens seeking participation in occupations across this nation. Denying citizens life, liberty, comradeship, and education in American military programs is unconstitutional. The equal protection laws have been one sided, not considering the rights of American citizens. America trains with foreign nations but denies its own citizens military training with educational benefits; and that's dishonorable.

The federal government rewriting words to undermine the contributions of others, is an ongoing tactic Americans suffer from so they aren't heard and the truth can be suppressed. One of the best ways to bless the people of a nation is to share blessings with everyone. For a very long time, military personnel, law enforcement, federal agents, and employees of the state and federal government have received blessings while citizens remain unblessed. Wealth cannot be one sided without reprisals. Historical records endorse that greed is one of the seven deadly sins but tolerance has been practiced by American citizens, nearly breaking the camel's back.

American citizens have been the victims of numerous terrorist attacks; and victims at the hands of police officers or soldiers murdering them. Physical, psychological, and spiritual injuries run deep in all races across America. Those who are wise will heed the warning that if American policies do not change the outcome will be a nation divided. Scripture states any nation divided will not stand, so it's wise for current leaders to listen or lose their power through they're own defiance not hearing the truth. Sometimes the harshest words addressing errors need to be said, even if these words create mixed feelings. America needs many Boston Tea Parties creating change because the problems are very deep, very unjust, and very hateful.

Preventing war is the attribute of true leaders who understand that averting conflict will help build nations rather than tear them down. Cities in nations take a very long time to establish, even hundreds or thousands of years. Wars destroy nations and cities in hours, leading to reduced productivity. As Americans, let's make the necessary adjustments to improve America and avert war. Ignoring problems only means people want this nation to fail. Replacing this nation with a new government made by the people is what's needed. Our actions are more powerful than our words, so these words are only effective if obeyed to complete a positive task.

The smartest men and women in the military cannot continually steal the dreams of American citizens through the invasion of privacy. The invasion of privacy fuels anger, which leads to strife and physical conflict. America has not established an honorable legacy in the records of numerous nations recording the thefts and abuses it has conducted. America's legacy written in stone around the world, points to America's national inability to socialize peacefully with multiple civilizations in just a few hundred years of existence. Many civilizations have records dating back thousands of years in world history, compared to an American 1492 vessel landing, and a 1776 declaration on the Americas. After six hundred years, native aborigines from these lands still aren't being honored, even after the 1776 signing of the Declaration of Independence. Nations that last the longest respect the privacy of their citizens and the property of others increasing the nation's prosperity because citizens obtain happiness.

America denying its citizens the right to property through invasions of privacy is unconstitutional according to the Fourteenth Amendment. America increasing prices on land to bankrupt poor people making them struggle like slaves and changing the laws of past generations obtaining free land shows how discriminate people are. An American government that changes laws constantly with an evolving strategy for racial self benefitting to disallow others to own property only reveals how selfish people are. Families given thousands of free acres of land in the past now extort great wealth from people struggling to survive and are only offered lots with high markups. America has plenty of extortion taking place and the politics are designed to operate for a period of time to benefit some and exclude others. If you cannot afford land and a home why aren't laws from the 1700s reenacted to have land races. I believe it's unfair some families get free land and others do not receive free land when they're all Americans. The federal government hoarding land and bullying others to not have land proves that hate is being practiced in Congress.

There's tons of desert land but no water available while the fertile good land is massively controlled and dominated to keep out entrepreneurs. The United States government needs to have an annual raffle or race to start providing free land to poor people who desire to be farmers or ranchers but can't afford the cost associated with real estate. It's not right that some get so much free and then the laws change so that others aren't given anything. Millions of Americans believe the laws makers in America are conducting a financial war to disenfranchise them; and to enrich themselves including their families. Nations are loved by they're people for respecting their right to create, have, and share from their own free will. Americans who seek to strip away the free will of American citizens with thefts through invasions of privacy will create discourse that leads to physical conflict. We witness Americans murdering Americans for minimal reasons. When thefts occur, this increases the hatred people generate destroying any peace in our society.

Practicing the work of excluding others from equal opportunities is an unjust discriminatory tactic used by American governmental officers to judge with hate. Many Americans are punishing repentant sinners and denying them equal opportunities. The measurement that America's government has given in judgment shall be used as a scale to measure out judgment upon the government. Destroying the dreams, goals, and livelihoods of sinners has been American policy for way too long by officers of the United States.

The United States Patriot Act is disrupting American lives, undermining the prosperity of civilians domestically and internationally. American laws must be changed, and we must not be beguiled by ideological thoughts or strategies to do the work of sin in disguise. Nothing is hidden from the eyes of Yahweh, and unrighteousness done in secret shall be made known in the light.

To increase national security, pass a federal law mandating gun safes for firearm owners who possess licenses. Before being permitted to own a gun, one must own a gun safe to secure the weapon; this should be part of the mandatory training to obtain a firearm license. We desire no more mass shootings in which the license holders lost track of their firearms, so the weapons ended up in violent hands, committing multiple murders. School shootings have to stop. If weapons arrive in the hands of non-firearm license holders, the gun owners are prosecutable for not securing their weapons. No gun safe in the home, no issuing of a firearm license.

As Americans, we all can afford gun safes if we can afford the guns. Gun safe manufacturers can help through an assistance program for the protection of communities, providing safes at discount to firearm owners, with the safes being tax deductible. Part of the firearm licensing will be a mandatory class on securing weapons and preventing theft of firearms. The licensing of firearms must contain a rule that guns must be secured at all times. The gun safe must comply with a strength and durability standard written into law so gun safes cannot be easily broken into. Secured firearms means fewer criminal shooting incidents and reduced potential for terrorists acquiring stolen weapons to attack Americans. National security is important!

Many adults with guns have been irresponsible because children, teenagers, and young adults have gotten access to weapons and committed horrendous mass murders in schools. The problem with school shootings resides with parents who are uneducated and poor parental guidance without good standards. Parenting without child guidance standards requires a new framework for establishment. We need new standards for modern educational systems as a mandatory curriculum. Parents must be taught how to parent. Teaching a parent how to parent can prevent so many crimes from being committed by neglected young minds who do not have good guidance from their parents.

Academic problems also reside with educators possessing credentials not befitting themselves. Psychology is being used improperly by academic junior high school and high school teachers. Educators are angering students by using psychological methods that do not work as intended. Many teachers are contributing to student psychotic episodes by triggering them to be enraged to the point of misbehaving, which leads to violence. Psychology is a systematic methodology to help people. Psychology is not designed to torment or disrupt lives to the point of generating anger. Psychology is deigned to help bring awareness to personal errors through an inward gradual self reflection. Helping patients become aware of the thoughts that need a practical solution to fix is partially how psychology works. What's happening in many households or schools have parents or teachers attempting to practice psychological techniques without being proficient surgeons in how the methods are used. It's very dangerous for non-psychologists to use psychology because of their lack of educational formidability. Psychology is supposed to used in a clinical office to establish a comfortable patient therapy session not enticing anger or aggression. Parents or teachers voicing what they want is agitation opposite than psychological techniques. Psychology uses an array of techniques through communication to bridge a patient to psychologist open conversation never meant to create an upset relationship.

Psychologists are trained experts who are familiar with all mental disorders and perform a psychoanalyst approach to determine the extent of the problem. Sometimes the problems don't originate with the patient but with the parent who consumes alcohol or uses illegal drugs. Psychologists examine many factors and not just the patients themselves who can be influenced by their environment. Psychologists can evaluate patients to determine if there are sociopathic or psychopathic tendencies for being criminal in behavior or adapting to one's survival in their environment. Psychologists initiate communication with patients who desire to engage them in dialogue. Forcing dialogue upon others isn't the way psychology works and some techniques once used in the past are no longer effective in this current generation. Psychology has changed just as medicine has changed. These days psychologists are professional listeners because young people as well as adults need to be heard as they deal with depression, addiction, and mental illness.

Psychologists are licensed professionals and are not parents or teachers who are unlicensed doctors attempting to treat patients outside a clinical setting. Parents cannot consider rearing a child the same as being a professional doctor. If parents or teacher are thinking of themselves as doctors well then we have a contributing factor to why school shootings are happening or why teenagers are being highly disruptive. Parents or teachers messing with a teenager's mind and playing games with their emotions or beliefs is dangerous. Also families with soldiers or veterans messing with a teenager's mind and playing games with their emotions or beliefs is dangerous.

Soldiers should not be taught psychology, for many of them are drug users and alcoholics with mental problems. Many soldiers refuse to acknowledge that they have mental problems and are living in a state of denial. Soldiers conceal their mental illness in many ways to maintain a level of manipulation over others. Many soldiers possess brain defects and abnormal brain growths in the cerebral cortex. Some soldiers have brain defects that can be attributed to physical trauma, and the strange behaviors or irrational thoughts they fantasize. Many soldiers have an obsessive-compulsive personality disorder in varying degrees. Soldiers awakening to nightmares, sweating in fear, or recollecting flashbacks of war create problems for many families. Some soldiers have none of these aforementioned post traumatic stress problems but they will want you to believe they do.

Many soldiers with mental disorders, want to lord themselves over others feeding their narcissistic disorder rooted in behavioral complexes they repress as they initiate conflict with others.

Active-duty soldiers and veterans frequently invading one another's personal space is a problem ignored by the US government. Some soldiers are kleptomaniacs who've repressed their urges for only treasured items of high value to victims. A veterans' mental disorders, like narcissism and obsessive-compulsive disorder, create hateful, disruptive disturbances within families. The destroying of another siblings' life because of all they're predatory behavior through thefts arrives in a conscienceless mind not caring while murdering the soul of victims. For the United States government to care even less than the conscienceless veteran stealing another person's life makes citizens feel one hundred percent hated by their administration. Nothing has been done by the Veterans Administration or law enforcement to correct the wrongs they've subjected others to by letting these psychotic veterans roam free. Veterans have consciously designed ways to inflict harm and strife on those they oppress. The truth is, we've all heard news stories in which someone has murdered all the members of their family. Some news stories are about veterans committing murder, and other times, it's victims doing the murdering to escape from the predatory veterans or domestic partners harming them.

With male siblings, these grotesque, thieving narcissistic behaviors can result in violence and premeditated attacks on the character of victims. The problem is women, children, and families are suffering because the narcissistic veteran is attacking their cognitive personalities through domestic violence and repetitive tormenting. For some veterans it can be the training and for other veterans with a mental disorder they just find enjoyment in being cruel to others. Sometimes being cruel to others finds its way into prisons, hospitals, and homes veterans reside at. Psychological abuse from active soldiers and veterans upon families is a grave problem in our American society.

Soldiers and veterans involved in substance abuse have been attacking the conscience of military families. Soldiers and veterans have been creating dysfunctional children who, in turn, attend schools being altered by the hatred. Poor parenting is a major contributor to problems in schools. Siblings who've trained in the military and have something wrong with their brains which was not detected at first this is a huge American problem disrupting normal society. The behavior within the military is reaching outside the military bases and into civilian households. Soldiers trained in military tactics are using methods used in war to oppress civilians and harm lives, often clandestinely. Civilians who could have been great role models now have destroyed lives and have been discredited, to the evil delight deviously minded siblings militarily trained. The American government has remained on the side lines and hasn't helped civilians with these problems created by delusional veterans.

The American problem with violence resides with educators not knowing how to communicate effectively with young minds who process incoming information differently. Some students are quick to understand incoming information. Other students may have a section of their brain growing slower or having less healthy nutrients so it's harder for them to process information as quickly. Students can also have medical conditions where the distribution of hormonal compounds is imbalanced and this effects their personality or development. The American government is ignoring the mental problems of veterans, and we have strife escalating to violence. Students who were once talented good pupils now have lives completely destroyed because of jealous military veterans who are psychotic siblings. Once the American government, its law enforcement, and the judicial branch labels a person badly a normal life can never be resumed with prosperity.

Many young people unknowingly are the victims of narcissists secretly disturbing their lives as well as their minds. Without looking at all the facts theAmerican government has ignored the victims of narcissists and has even aided the narcissistic vendetta to discrediting victims.

People in government positions have been so unfair with civilians struggling hard to be uplifted to a higher status in life. Problems can only be fixed when someone is willing to expose the perpetrators so they're unable to conceal their criminality. When someone's a criminal, they look for any place to hide their devious behavior, and especially when they're harming others. In business the rule is location, location, location, but for criminals it's the same business structure minus all the laws which they refuse to obey.

Educators must take responsibility for these violent problems in schools because they're not conducting themselves properly to address issues with mentally ill students. Delegating that students see someone but they're not informed its a psychologist is invaluable if they're told its only a counselor or a teacher. Professional psychologists don't want to alarm students who are already scared, disturbed, or defensive especially if there are other issues taking place in the home. Educators are failing their duties assigned to them to instruct young minds because numerous factors each effecting students differently. The lack of constructive extracurricular social groups and activities in schools is alienating students who have too much time alone. Students who are alone too often begin contemplating negative feelings, possibly brought on by bad parents or militant families. Educators need more insight into the households in neighborhoods where our students reside to prevent neglect, abuses, and lack of basic resources like nutrition or medical care.

Violent video games with strong subliminal messages are desensitizing students' feelings, thus reducing empathy, which contributes to increased aggression. Violent video games contribute to thought errors in some students who have nothing else but the video games if the parents aren't around giving them guidance or nurturing parental interaction. Some students are alone at home and because of this they develop an increasing level of antisocial behavior living in a dysfunctional household. The lifestyles students develop often originate from environments they're exposed to which can lead to concealed feelings, depression, anger, and even sociopathic or psychotic episodes. In some neighborhoods gangs may be so oppressive that murdering another human being maybe the only way to escape being killed for young students predatorily victimized. The psychologist would be able to discover if a student is sociopathic or psychopathic and if there are other factors disturbing the teenagers behavior. Young students who murder have other underlining problems that need addressing through compassionate psychologists. Maybe all schools must make it mandatory that students must see the psychologist once a week for an hour. All students seeing a psychologist is better than another mass shooting or violent attack happening in our communities. Antisocial student behaviors lead to depression which contribute to suicidal thoughts. Stopping the progression of students wanting a shootout to end their lives is what we must help to alleviate. Students experimenting with drugs or alcohol have increased the social problems in their lives. Drugs and alcohol create a combination of developmental problems to school engagements with other students. Students who don't fit in or are treated as druggies often get ignored and outcasted. Some students only know about drugs because their parents are users exposing them early on to narcotics, smoking, and alcohol use. Need I remind America's people that "In God We Trust" is on many American documents as well as our currency? Either Americans obey Holy Scripture, heeding Yahweh not to covet anything that is their neighbor's and not to steal, or Americans do not desire to heed Yahweh's commands.

Not obeying Yahweh's hand; is directly giving favoritism to Satan's hand. Not obeying the voice of holy messengers who are modern-day saints and priests is one way of missing the opportunity to obey holy messengers coming from Yahweh the almighty creator. Not one government on Earth will endure when it trusts in its own power. Beware of human beings who place human ranks above Almighty Yahweh's uplifting power that has endorsed angelic authority.

America has been ignoring the Holy Spirit working through saintly lives much like Moses, being a holy messenger for Yahweh's Commandments.

America's government and its employees who don't believe wise messengers sent to guide a nation to obedience to Yahweh are in opposition to heaven's authority over the earth. Many saints, nuns, priests, monks, angels, and popes agree with my words for Americans to start correcting themselves through sacramental confession in reconciliation. It's very difficult to guide people away from impending doom when they refuse to hear your voice and spend their lives discrediting your efforts. The Holy Bible says you will know them by their fruits. Many within our American government walk in sin through wide paths, so beware of the outcome. America has challenged the power of ecclesiastical ranks, thus challenging holy judgments supported by Almighty Yahweh.

The American reality is that hundreds of school shootings have resulted in many murders, so let's reeducate our students with honorable military guidance. In some states, we can see billboards advertising military academies and guided instruction. Many American families have parents who aren't educated enough to raise children, and military academies would suit these children very well. A positive connection equals happiness for youths and adults desiring productive lives. Being separated or disconnected in communities will increase sadness in students making more occurrences when we witness violence in our society.

The strongest men and women believe they're untouchable, trusting in their own strength, and they cannot live every day on their guard without just one moment at rest. In many nations, the strongest men are gunned down, beheaded, and bombed frequently, so it's naive to trust in one's own strength, which is false. Special forces officers, active-duty soldiers, and veterans who put themselves in an untouchable place as if no one can overcome them have been deceiving themselves. World history is filled with strong men and women toppled in their arrogance. Many strong warriors in history were toppled by the least likely person to dominate over them, even in biblical times. In an instant, empires can be removed from history; for even pandemics of ancient times ended long eras for wealthy nations wanting to rule over others. To strengthen a nation is to train everyone, for everyone could become a victim. The victims of terrorist attacks are often citizens being murdered. Who's more dangerous: the terrorists or the government that offers its untrained citizens up to be murdered by the terrorists like weakened lambs sent before lions in a slaughter?

Terrorists don't care if a soldier can lift five hundred pounds or a thousand pounds, for a knife, bullet, or bomb ends the life of the strongest men. I'm making these points to emphasize errors in the thought process of Americans who exclude their fellow Americans. Powerful strong people are just as vulnerable to terrorist attacks as are smaller weaker people. Americans who spend time weakening their own nation by denying their own civilians' requests for privileges are betraying their fellow Americans. The American government is creating castes, placing itself above citizens, and believing it possesses all authority to deny humanity the right to excel in any desired occupation. Who are these people being so judgmental of others and not seeing themselves as sinners betraying the body of Jesus Christ and warring against the kingdom of heaven? These people are not our brothers and sisters in the kingdom of heaven, but they're our eternal enemies betraying our lives with their arrogance and pride hating us.

Something is wrong with the intelligence of many in the highest seats of the American administration. I'm far from being a person who refuses authority, but when people want to be treated as authority without being honest, or holy I see through the bull feces. Yes, I'm saying mental illness does reside in people elected to state and federal government. America has problems in its judicial seats, legislative facilities, military departments, and law enforcement offices tied to political houses which transgress against Yahweh.

Billions of people around the world will agree with my words about mental errors in American government. This is pretty significant when billions of people globally agree on something and that something is wrong with Americans. Even Americans know something is wrong within America because we witness national protests being seen on the international news. Americans in government are quick to ignore important issues that are political and to act like dirt or grime can just be covered or misdirected so it's not noticed. In the General Assembly of the United Nations world leaders know that corruption exists in America and it's being covering up so they laugh at our American leadership. In international meetings of high importance, America is thought of as the mentally ill child unable to be tamed, having lost its nobility. Yahweh has heard the many prayers and supplication from the faithful believers regarding injustices committed by Americans that have risen up into the heavens. We are not superior; we are failures who need to make improvements on many tiers of our social, political, economic, and religious structure.

We've addressed the problem here; so now, we need to correct the problems in government because the old ways of thinking are outdated by our modern technologies. We communicate with nations around the world now lets fix our problems with nations around the world. Terrorists don't care if potential targets have military service stripes on their sleeves or no stripes at all. Why are the hearts of American leaders so adamant against strengthening American citizens? Have American citizens done something wrong by desiring to improve themselves and get a better education?

No American citizens have done nothing wrong; and yes, America's leaders have done something wrong. They've hated their own citizens through inactions and denials. Millions of Americans feel hated by their government leaders, striving to inflict harm upon them. America's government has created a realm where citizens lack accomplishments and the pursuits they have are met by evil hearts warring against their desire to be happy. The attitude in the American government that it's only for us; and we're going to deny you with hatred till you go away has to be uprooted. Those types of hateful discriminately evil people have no place in government with such a negative attitude towards citizens in this nation in which they live. Are these people giving denials to other American citizens even Americans or are they enemies of this nation? All forms of hatred must come to an end, for hate is of the devil's kingdom and not of heaven's kingdom, where peace resides. The outcome of the hatred practiced by members of the American government will cause another civil war in America, a prophecy I can see unfolding. Wise men seek to avoid war; and foolish men transgress their ways to start wars. It has been said since the times of old that one should listen to those with wisdom and see the signs that can be interpreted of the things to come.

On another note, government should make it a policy that all allegations against someone who runs for public office must be brought forward before any election campaign. Once in office, criminal allegations have no validity, tying up too much time for politicians, who need to focus on greater issues of importance like running a nation and improving its national security. Legal cases create major conflicts of interest for elected officials, hindering them from performing their duties to the best of their ability.

New statutes must be established to prevent criminal allegations against elected officials after they've entered their elected office. This policy change prevents allegations of corruption, illegal activities, and scandals from consuming the public media with inaccurate information. Issues from election tampering to men or women making sexual accusations disrupt public news during complicated political issues concerning this nation. A new legal statute must be established so the American people aren't tied up with allegations made to discredit public figures running for office. American politics shouldn't revolve around what may have happened ten, twenty, thirty, or even forty years ago in a politician's life. Jealousy causes people to do many cruel things for fame or fortune in America. Some elected offices, cabinet positions, and government positions must have greater exclusion from repetitive false claims that were not brought to the attention of the public many years earlier.

Attacking public servants only after offices are assumed and oaths are taken is a strategy meant only to discredit those elected. Not bringing forth allegations in advance of elections means a new statute needs to be enacted to prevent these attacks after elections have taken place and new leaders are in positions of authority. During national elections, any allegations or tactics to disrupt the work of politicians running for a office must be halted. Allegations because of jealousy or the potential for a higher financial payout must be restrained and stopped to prevent false crimes being committed against politicians. Discrediting people can be an activity foreign governments partake in to disrupt voting and tamper with election turnouts. The criminal network business of organized crime syndicates can use allegations as a means of bribery to extort money from elected officials. If a politician mistakenly dated a woman for example many years earlier not knowing she was affiliated with crime syndicates this event could have been preplanned to sabotage the politician career while he unknowingly had fallen into a spider's web. There needs to be a law created to prevent elected leaders from being victimized by allegations designed to discredit them during or after political elections.

Americans continuously practicing divide-and-conquer strategies on the people of this world needs to stop because we're all one people living together in this world that needs us caring for it. Dividing our world isn't the answer. We should be focused on uniting our world as the earth's population increases. Transportation systems will continue to improve worldwide. Traveling from one country to the next needs to become a normal everyday activity for all humanity so we grow in our social relationships.

America must stop doing the work of dividing other nations or it will guarantee that Americans establish a place of resentment in the hearts of billions of people rejecting it. Those with mental illness who want dominance as a tactical advantage over others haven't come to understand that most of humanity seeks peace on Earth. Constantly living with the war mentality is exhausting and there's more to life than that.

Americans need to recognize all humanity as one blessed family on Earth. The universe is wide open to be shared by humanity that lives in a righteous way. The devil has inserted himself into the hearts of mankind to battle over this planet when there are so many other planets for adventurous habitation. Taking advantage of others in every facet isn't an honorable way of living it's the evil path that lusts after ill-gotten riches rooted to greed that harms others.

Division in spirit is not a way to build a nation or unite a world with common values. We should love one another, a commandment any nation should uphold or find itself divided from all humanity. Micromanaging continual war against others will only bring destruction upon those refusing to seek peace. We can train all Americans to protect themselves, and then we can seek a peaceful living with neighbors, pursuing technological advances for humanity's benefit.

I support the Second Amendment for the right to bear arms. I was raised owning a firearm, and I knew people who had hobbies involving firearms. Some of my friends were hunters and gathered meals for their families. Other friends enjoyed the sport of target shooting and the hobby of inventing new firearm products for sale. Other friends lived on rural farms or ranches, so a firearm was added protection, being far away from cities. I've had friends who felt safer on a lonely hiking trail with a firearm, walking all alone, not sure if wildlife would confront them or if someone would be looking to harm them.

I will not make any attempt to take away the firearms of the American people. I'll be honest: many Americans need a better training course and a stronger, more intensive psychological evaluation before owning firearms because they require a higher degree of responsibility. Some people deserve to have firearms and don't have them, while others have firearms and don't deserve them. How do we achieve a balance for firearms? We make everyone go through a rigorous course of safety training in gun use and a psychological screening. We need a national gun safety standard and one or two years of mandatory military training for all Americans. All this training will also require weekend drill training for all Americans periodically to update their files.

The American government has separated itself from religion. The National Emergency Security Task Force and the National Emergency Strategic Task Force are two groups with a religious moral code that follows sainthood. NEST was created around the belief that the crown of thorns Jesus Christ wore at the crucifixion held all the peaceful nature of the universe because Christ was tolerant even onto death.

In Scripture, thorns can choke others with unquenchable thirst and thorns can represent demonic people who are like dried weeds that choke living people and weeds must be gathered together to be thrown into the fire. Thorns can also provide protection for an eagle's offspring. Thorns can also be used to protect us from evil thoughts for Jesus Christ is also our protector and he wears the crown of thorns. Thorns are symbolic for Catholics and Christians to live a fruitful life like blossoming roses. Roses are symbolic as being handed to the Virgin Mary who also prays for our protection against dangerous thorns which are ungodly. Jesus Christ overcame death, and his fountain of life-giving water restores our soul to fruitfulness against any ungodly thorns attempting to choke life from us.

Thorns are also believed to symbolize sin ensnaring people like the claws of an ungodly beast, digging deep into the flesh and bringing about harm that results in death. Thorns cause stinging pain and can release toxins, which create more discomfort. It's said that the claws of an ungodly beast carry toxins that can cause infection and death, so thorns symbolize these claws. The tree of the knowledge of good and evil from the Torah also was also in the presence of thorns within the Garden of Eden. We are to tend our gardens, removing any thorns which are sinfully evil thoughts or desires, striving to entangle us and drag us down into the depths of Hades. The Holy Bible teaches us that true believers are one with Jesus Christ, so wear the painful crown of thorns our Lord also wore. NEST is a nongovernmental organization that needs to be established to create balance in America without permitting the federal government to strip away all power from the people through rebellious denials.

The American people are hardworking, and paying their taxes being willing to do any job to keep this nation moving forward. The American people are resilient, overcoming hard times in pandemics, wildfires, floods, hurricanes, tornadoes, earthquakes, storms, riots, terrorism, mass shootings, and financial distress. Americans do their best to put together the pieces of their lives, going through many hard times, often going to other states to work in an attempt to support their families. Americans have been passing away in hospitals because of an infectious disease, and Americans will travel to go see those they love even when its in difficult times.

All Americans are unique and have their own voice, which matters greatly in this nation where they live. Past and present generations of Americans have fought for their nation, and many Americans lost ancestors due to war within the Americas. Americans have supported many of their leaders from the state and federal government only to be betrayed.

So why are Americans, who give so much to their nation, not understood and given the opportunity to vote outside their own state? All Americans are dealing with different circumstances in these uncertain times. Many American citizens aren't financially able to get back to their own home states to vote in the presidential election. Provisional ballots in states other than the home state of the voter threaten federal fines to those who desire to vote. The provisional ballots in all fifty states aren't designed to accommodate and accept votes from out-of-state Americans when they should be. What makes matters worse is that millions of Americans who desire to vote arrive at a polling location in person and don't have a booth set up for out-of-state voters while possessing a valid photo identification.

Millions of American voices are being silenced by the very people supported by their tax money. In American national elections, many Americans can become arrogant, treating other Americans like their votes don't matter, and will not be counted, or will be discarded. Millions of American voices are being treated by other Americans as if their unique words have zero validity and they have no constitutional right to vote. A condescending behavior is the position many people in government offices have taken toward the American people when it comes to election time and voting. We can unite and fix the current errors by initializing the United Security Act.

We need a law so that all states have a booth for out-of-state voters who couldn't make it back to their home states for whatever reason. We also need a law so that there is a section on the election ballot for out-of-state voters to vote in the national election, because their voices matter. The fifty United States and their territories must implement these laws to enable millions of Americans not to be defrauded out of being able to vote. America needs justice to help prevent national violence among the people.

The US federal government must make these complaints law because all Americans matter. I doubt many Americans would be willing to take the risk of trying to vote twice when they must enter a voting booth in person on camera using a valid state picture identification while they're completing an election ballot. All states are obligated to provide justice by providing American citizens the opportunity to vote no matter where they're physically at. A picture identification only shows a person's locale where they return occasionally. There are millions of commercial semi-truck drivers, for example, living on the road and working to keep all

Americans fed. The semitruck drivers are moving medical supplies to hospitals and working to keep the economy running during this pandemic. All these truckers' and their millions of votes matter, in addition to all the other Americans working outside their home states and dealing with many unfortunate circumstances.

2

HONORING STUDENTS GLOBALLY

In the United States, four-year universities typically host Greek fraternities and sororities. Globally, there are many students pursuing academic certificates in programs continuing their education who are not permitted to belong to a fraternity or sorority. Violence on campuses increases because there are fewer opportunities to be accepted socially and academically among students enrolled in education programs. Troubled students in high schools, colleges, trade schools, and universities don't need rejection while they pursue education improvements for their lives. The transference of bullying or mistreatment beyond junior high school has no place in our academic society.

Stereotypical behavior to belittle or put down others without a four-year degree or graduate degree has contributed to strife on campuses, which needs to stop! Men or women psychologically tormenting their peers repetitively is a behavior of adolescence and not a behavior of academic adults contributing to society. Students intentionally participating in exclusionary tactics and bias rejection towards fellow students through alienation is unhealthy. Students placing castes on others who pursue academic educational certificates is unfair and prejudicial.

Students who've made the decision to improve themselves beyond high school, should be allowed to wear graduation gowns, to honor them. Students surviving in this world from high school becoming adults have earned the privilege to wear graduation gowns for any educational program after K-12. The practice of denying caps and gowns to students after high school who are advancing themselves is dishonoring them. Aren't we supposed to build up our society and make students feel proud about pursuing academic excellence? We should honor students who've made the righteous decision to better themselves. Not acknowledging students' advancements is a terrible way to encourage education, especially when students influence others around themselves.

Students in life treat people better when they're being honored for their educational progression. Universities have denied certificate students caps and gowns, even though these students have made the right choice to become active contributors to society. When students complete a study course and they're treated by others as though they've accomplished nothing of merit, then this behavior doesn't foster positive growth. Students continuing their education are doing the right thing.

Student excellence should be rewarded because in today's technological world, the most minimal improvements can change world markets and science in numerous ways. Certificate students shouldn't be excluded and treated as less important than four-year university students.

Certificate students shouldn't be denied the privilege of participating in fraternities and sororities. Behaving in a way that is acceptable to society is important for new university groups fostering socialization between students. Academic institutions are crucial to all nations of this world and we must guarantee society that no harmful psychology is being practiced on students.

Psychology is a field of study that cares for others who've been harmed or need to recognize their own errors in thought for a self correction without inducing aggression. Using psychological methods to disturb minds or harm others endeavors is unprofessional because not everyone thinks or acts identically. Trying to force others to conform to the will of those living in autonomy is harmful to the character of victims.

This no-harm psychology policy must be put in place for military institutions because behaviors originating from these branches of service disrupt the civilian society. Maintaining peace in our society must be a mandatory goal. The disrupting of our society on many different educational levels isn't how psychology should be used. The attacking of faiths using mentally ill narcissists who were former soldiers is unprofessional.

The maintaining of peace on university campuses is mandatory as is the stopping of criminalistic tactics implemented by fraternities, sororities, and military personnel conducting home invasions. The invasions of privacy as a prerequisite strategy to obtain wealth illegally on campuses must stop! Saints have the sanctity and the Holy Spirit to overpower the strongest of demons.

Government officers with a mental complex have sought superiority to reign over the holy sanctity of faithful believers in Jesus Christ. Many of these faithful believers in Jesus possess ecclesiastical offices, entitling them to dominions others cannot possess because they're not being holy. Like Judas Iscariot who stole from the church there are many with identical ways who also reside in the United States government. Sinners who think they always supersede others have a desire to reign over what they're not entitled to have. Americans who violate the sovereignty of Catholics are violating Yahweh's decisions to bless the faithful and not the unfaithful. Demonic people want what angels and saints have because they're cut off from entering heaven.

The highest positions of faith are eternally unattainable to any person not living as a true saint. Many demonic people obey their own desire while opposing Yahweh's orders.

The saints and angels have the spiritual holiness and authority to make judgments over humanity. The saints and angels are doctors over humanity having souls greater than sinners.

Property stolen from saintly students by those hands which sin and are unholy can bring an end to this world. I trust the Saints and Angels with all the knowledge of war but I don't trust sinners who serve government rejecting Yahweh's judgments with their spirit. Often Yahweh grants power to those individuals worthy but whom many people in society reject for leadership or rule. The Saints and Angels are royal to Yahweh living fruitful having a deeper understanding of all life than humanity. Being a nuclear physicist for example is minimal to a Saint and Angel with greater knowledge in this subject than humanity. Saints and Angels don't think as humanity thinks. The Saints and Angels have the heavenly power to destroy humanity but they don't desire to destroy the world Yahweh has created.

Many narcissists in world history have caused wars upon this planet which have caused millions of human lives to be killed in battle. Mental illness is extremely dangerous to the well being of others because psychotic minds cannot stop their obsession. A mentally ill person will always smile at circumventing the laws of Yahweh thinking their thoughts grant them higher status within heaven than others. Demonic people with mental illness can never accept that they're living completely opposite to the way Yahweh desires. Many church leaders aren't living the way Yahweh desires them to live, thereby creating an atmosphere mentally ill parishioners flourish within victimizing others secretly. Universities need to provide safe environments where mentally ill individuals cannot steal from or harm holy students. Some students write to discover the things to come hidden within Holy Scriptures to uncover the future in their study of theology. How Yahweh intervenes in the lives of faithful servants is different for each person given talents.

Narcissistic government personnel have not obeyed heaven's first law, to love Yahweh which requires obedience to the Holy Commandments given to Moses. The US government is opposing the sanctity of students. Americans have accomplished millions of abortions, mass murders, and the seven deadly sins through their disobedience. Help students become better students by encouraging the right deeds. Work together to oppose those who are doing evil deeds.

Those who live holy, and religious lives can in no way be compared to Americans who sin living in worldly acceptance. Jesus Christ was rejected and not accepted by many people so you can except the same when you live a holy reverence to inherit eternal life in heaven. The US government has passed laws that help enable the predatory behavior of criminals to thrive in violating Holy people's lives. In the end times there shall be more injustices piling up from the US government against all righteousness in an attempt to approve themselves above heaven. Americans have sought dominance over all things possessed by ecclesiastical offices, to oppose heavenly laws governing who receives different spiritual gifts. For example the talent Yahweh gives you isn't the same identical talent that Yahweh would give me. You have a different reverence than I do, and I have a different covenant with Yahweh than you do. All human beings are unique and to treat them as identical isn't a just judgment. Those who sin wanting what they're not blessed with are those wanting to wear the crowns of life they've haven't been awarded. Evil in the hearts of many propels them to continue their transgressions unwilling to surrender themselves one hundred percent to Yahweh's commandments relinquishing what they've committed through sin. Hold onto what was stolen and never enter heaven or conceal and hold onto what was stolen violating Yahweh's eternal commandments? Many have chosen to continue hiding what they've stolen violating hearts and souls so that they continue violating Yahweh's laws. Judgment day will be a great day when all the lying and deceptive thieves concealing their evil deeds will be exposed for the entire universe to witness the falsehood they keep within themselves. Where falsehood resides very near is the Devil who's also false exalting himself to be above those he can never be above.

Many Americans belong to the beast government spoken of in the Holy book of Revelation, through their warring with heaven's sanctity of rule over the earth. All nonbelievers not accepting that America has become the seat of the Antichrist are deceived. Satan desires only the most powerful seat to corrupt souls on the earth. Yes America is powerful and it's leaders kill life at will for this is what the Devil wants and Yahweh says do not kill. Only the Saints and Angels are permitted to kill. Saints and Angels use the power of Yahweh to kill so that they cannot be judged as being unrighteous. Saints and Angels are patient having ways that humanity doesn't possess. Humanity desires to use all weapons of mass destruction while Saints and Angels have no such way. The teeth and jaws of the ungodly shall be broken for desiring to eat what Angels or Saints eat. Those in the bottomless pit aren't permitted to eat what the Angels and Saints eat.

The devil was a murderer from the beginning, and he loves all the murder taking place in America, where his demons are influencing millions of people to sin.

The question to ask yourself is where would the Antichrist who opposes Jesus Christ seek to have his seat? Remember that Satan wanted to be more powerful than Almighty Yahweh. Would the Antichrist want his seat in the poorest nation on earth or would the Antichrist want the most wealthiest nation on Earth to be his seat of power? America has millions of sinners living like those who were in Sodom and Gomorrah before it's destruction? Americans cannot blatantly lie, saying multimedia sexual pornography doesn't originate from Americans being distributed worldwide online. Pornographies sin exists in stores, American homes, and on the internet traveling from America to other cultures impaling thorns of sin into marriages causing couples to be unfruitful corrupting many nations. America endorses the murder of fetuses with the laws established by the Supreme Court. America has influenced other nations with it's sin to copy the murdering of fetuses through abortion. In my judgment I find the Supreme Court unrighteous and beneath the heavenly judgment of Saintly Angels. Why hold in high esteem sinners in government who are eternally beneath you while your living more reverently righteous than they are? You stand closer to Yahweh's holy throne so why are you subjecting yourself to those who are corrupt in government undermining your holiness? Why did I speak about all these problems, because it's the responsibility of parents and teachers to educate students in the right paths they should take in life. Yes speaking this way to tell the truth can be considered callous or less sensitive but I'd rather be honest in my approach to addressing errors in our society. We need to get tougher on our students by first getting tougher on our academic administrations to improve their methods of discipline. We don't want people going to Hades obtaining a life in torment. Being tough on people who need guidance is necessary to save lives and prevent students from being corrupted by societies way of accepting sin as common place.

Ungodly demonic spirits are drawn closer to those committing sins, and with so much corruption in America, the demons visiting this nation are in the billions. In the Holy Bible, Jesus Christ asked a demon its name; the demon responded its name was Legion for there were many. Since there are many demons from ancient times like the demon named Legion, we're seeing chaos in America. Americans are becoming more corrupt every day, except for Saintly people living holy and obeying the Holy Bible. We all hope sinners turn to Jesus Christ Our Lord accepting him and believing in the him while studying Holy Scriptures from the Holy Bible.

Anyone reading my words can't deny Americans are murdering one another and that millions of criminals thrive free in our communities. US soldiers and veterans have become criminals that repetitively steal and have escaped punishment for violating laws through numerous jurisdictions. Members of the American government have thrived off sinning to elevate themselves in the realm held by the Devil while they've been unprofessional towards citizens. American soldiers and veterans believe it's their right to invade anyones privacy. I expect to be discredited by those in government. Governments use a discrediting strategy all nations for they're avoidance of the truth, especially when they're been found guilty of sinning. Students have the right to not be treated with a total disenfranchising tactic like the American government has used upon many of its citizens. The police, FBI, and the federal government as a union doesn't want to see students or felons who've changed become wealthy entrepreneurs. So many laws created in the United States aim for one thing to label others badly and to keep that labeling indefinite so prosperity is diminished humongous in the lives of citizens. Covering up discrimination and hatred to undermine others has been practiced by federal law enforcement in the United States becoming no different than socialist or communist countries. If your one of the millions of Americans who've experienced the oppression then you know what I'm talking about. America has been built on committing crimes and covering them up with numerous generations acting like nothing is wrong because they've gotten away with it for many years. There's one fact that all those who've experienced this hatred can look forward to is that they cannot escape getting away with it in the end times when Yahweh judges them. My kingdom is not of this world and as we get older we understand that dying isn't far away in the time we've been given. I'm confident Angels will destroy the history and legacy of those who've put so much effort into destroying our legacy. Before the Great Flood wiped out humanity except for Noah and his family the inhabitants of this world only desired their seed to flourish along with their legacy. So as many Americans oppose holy people Yahweh has forgiven be confident that they're not favored or going to possess and eternal legacy like you who live according to the way Jesus Christ summoned you to live. Don't ever allow the unholy to ever think they can dictate how your life should be or what your career or fortune should be. Yahweh has counted the sins of those opposing you and as long as your life is changing for Jesus Christ to become more fruitful you have a legacy to pass onto your children who are the students.

Thieves throughout history have started wars all because their invasions upon another person's privacy. Governments constantly practicing tactical theft operations against students through trained soldiers is offensive. Patriot laws passed to invoke oppressive sinful transgressions upon people will be the catalyst guarantee igniting future wars to dismantle America's government. Disguising crime with names like the Patriot Act used in ways outside the facilitation to gather terrorism intelligence is how America uses these data mining programs. America has been practicing espionage on Americans because its profitable. The American government has been enriched with trillions of dollars of wealth all through illegal wars being conducted against the citizens of the United States who have intelligence. Nations that seek peace have prosperity, but those that offend Almighty Yahweh reap wrath upon themselves. Many people will not like what I've said but what do you think the Angels of heaven think when they look down from heaven witnessing the chaos across America. If you don't like what I've said well then change all the errors in government that continue to harm lives disenfranchising American families. Those who aren't willing to change the problems within government shouldn't be anywhere near the government. The Civil Rights Movement hasn't ceased it's ongoing and its still experiencing corruption from within government. We need a second round of Civil Rights Movements nationwide because minorities still are being mistreated and victimized. There needs to be a new bill passed for Civil Rights equality repayment through both houses in the American government by through this generation of brilliant students becoming politicians. This generation of students will replace the old political seats held for multiple terms in the American government by getting rid of old ways of thinking corrupting this nation. Many people who haven't been paid for multiple generations being disenfranchised need to be paid. External investigations need to be brought forth to expose what the American government has stolen and to reduce this wealth from the government giving it to the original authors and inventors or reap the repercussions. A nation of millions of disenfranchised people have a voice and to not pay what's owned means that the American government is no longer a democracy but a government covering up its crimes for financial gain. In the past we would expect tyranny to come forth from third world nations but not from America. Lets see what happens for our students, families, and the victims who've acknowledged crimes against themselves. All the periods in America's past where technology originated from the original documents authored by others and information was stolen from victims incorporated into the United States government without victims being paid. I'm sure Americans expect hundreds of pallets of gold to be allocated for repayment to them especially when we have so many multi-billionaires being enriched while victims haven't been paid a cent. The United States government owes the America people not the other way around.

America is filled with millions of people who do not fear Yahweh Our God, trusting in their own intelligence, resources, and might. It's impossible to support any beguiling nation doing wickedness in disguise. The question to ask is will America coverup all its crimes and be like corrupt corporations that stoop so low as to burn documents or falsify information to prevent lawsuits? Will the American government refuse to investigate crimes it has committed because of its own racism, discrimination, prejudice, and bias hatred against minorities or felons its wrongly labeled? Students need to be taught of the truth since they're the generation going to invoke change in the United States. A repentant nation shall be given mercy and grace, but a nation that makes allegiance to Satan through its own actions is an abomination to Yahweh.

Many Americans have made their allegiance to Satan while serving the US government. Wickedness is flourishing in American society hidden behind closed doors and sometimes out in the open blatantly within our communities. Many Americans have denied participation in pornographic sexual promiscuity, adultery, fornication, polygamy, fraud, thefts, murders, rapes, hateful racism, oppression, gluttony, and millions of abortions refusing to do righteousness. Americans' live with overwhelming pride making themselves believe they cannot be overpowered by heaven as they trust in the military. Soon a Biblical catastrophic event will come much worse than a pandemic and either humanity changes so it educates students the right way by honoring them or the seeds of strife will be sown.

Satan is overjoyed with America's unholiness transforming its people into demons who move throughout the earth doing all manner of sinning for the world's nations' to observe. Satan is ecstatic that elected officials in America will not control their sexual urges, while nations around the world showcase our politicians corruption through multimedia empires.

America's political leaders living in sin have encouraged others worldwide to partake in deviant behavior and minimize unfruitful activities. Whatever happened to political leaders living upright, decent lives being faithful to their wives, churches, nation, and families? Just like muscles which are tore down to build them back up, sometimes we need to do this with poor leadership so they begin correcting themselves.

Many Americans have become the embodiment of sin. Many people oppose holy sanctity by refusing to obey the six hundred and thirteen commandments Yahweh gave Moses. Many Americans deny the existence of Yahweh remaining as atheists in disbelief. In the holy gospels, severe weather was calmed by Jesus Christ. In the Torah, severe weather was created for the destruction of mankind by the Creator judging humanity as unfruitful.

America has many atheists who've manipulated their way to top occupations. Even hospital positions have been compromised by unholy minions serving the Antichrist carrying disbelief. America has refused to correct itself, even after being chastised by Yahweh to change. Many governments deny the existence and power of the Creator even when disastrous weather envelops sinful communities.

Human beings are destroying the earth, so Yahweh is destroying human beings for destroying the earth. The first job in the Garden of Eden was to tend to the garden. Governments have irresponsibly destroyed the world, poisoning so many different forms of life and ruining the earth's beauty. So many leaders have failed to correct their nations' people concerning environmental emergencies.

I'm not anti-American. I believe that voicing a problem is one of the only ways to correct repetitive failures to begin doing what's right. I'm all for correcting America. I'm one of the few voices willing to say Americans aren't behaving like the best nation on earth.

Many nations have freedom just like America without having all the legacy of war. In the eyes of the Creator, there are nations greater than America. The Vatican helps many nations worldwide not being involved in wars. The Vatican sends missionaries through acts of love, charity, faith, and generosity. America can learn from the Vatican.

America's history proves that frequently university graduates have committed crimes while possessing advanced degrees. What about the certificate students who don't possess a four-year degree and were raised in dysfunctional families who didn't commit crimes? Students shouldn't be treated as being less important because they don't possess a four-year degree. A university faculty that makes students feel like they are less important than their counterparts is starting off on the wrong foot in a global society needing peace.

Making students feel under educated and like they cannot belong, isn't a message that endorses "No Child Left Behind." Students need social acceptance. A nation that doesn't accept its own people socially is divided at it's core foundation being, at odds with its own prosperity.

In America we have suicide, depression, gender inequality, homosexuality, LGBTQIA sexuality, immigration, racism, and anger dwelling in the hearts of many students. It's important to make our educational institutions peaceful, healing sanctuaries of safety, not places of division. Today, psychology is being used to do more harm than good, instigating conflict repetitively, to the point its offensive toward's students.

Some people have the fortitude to handle the comments of others without becoming overly affected by the words. Other people cannot handle harsh words without becoming overly affected. In America, some soldiers with knowledge of many different occupations, cannot handle psychological conflict. Soldiers committing suicide just proves that these individuals should have never sought the occupations they were seeking. In a wartime situation, how many American soldiers, if caught by the enemy, would fold under pressure just because of the words used? Many homosexuals want to be warriors but they cannot handle criticism, verbal conflict, physical conflict, and psychological trauma resulting in their suicide. The delusion many homosexuals perpetuate is that they're full contact warriors having never fought a physical battle but stealing from those warriors who've they've been impersonating. Many homosexuals want to get up close and personal with other men physically wrestling with them sexually but they're not full contact fighters. Many homosexuals have gotten into mixed martial arts to wrestle with each other up close and personal. Many homosexuals have gotten their way into commanding ranks in the military often because they've stolen information from valuable assets of the government compromising secret offices.

My point is that students need to be taught what they're getting into so when they make a decision they'll not regret their focused energy for a chosen occupation. Many universities offer military training, and some students desire to pursue a military career, being part of the LGBTQIA community. If someone cannot handle psychological torment, maybe military training isn't for them, for wartime battlefields are cruel, emotionless places where combatant feelings are ignored.

Many veterans are cruel having intentionally thought out plans, intended to inflict harm upon a student's mental process. Attacking the psyche of students to generate non positive emotional responses is how many veterans are using psychology. Many veterans with mental illness repetitively increase the intensity of their personal attacks on students knowingly practicing this behavior which warrants prosecution. Veterans that are overly critical and judgmental toward students can be very dangerous.

Some narcissistic faculty members holding authoritarian positions in universities who've hid their mental illness cannot accept a student's refusal to acknowledge their super-ego god complex. Many students have been taken advantage of because of their brilliance while narcissists on campuses rob them of financial prosperity. Narcissists conceal themselves on university campuses constantly thinking that students should be serving them and all their heart's desire.

Narcissistic thieves who hide in the ungodly realm that they've concealed stolen property taken from victims is where the demons reside. Those victimized don't even know they've been stolen from because they cannot often remember every detail they created on paper, or computers to aid their remembrance. Thieves have repetitively stolen the tools that authors and entrepreneurs use to aid themselves and their remembrance. Thieves have stolen many brilliant ideas worth billions of dollars being tons of money from authors and entrepreneurs victimized repetitively. This aforementioned reason is why narcissistic thieves are so dangerous because they leech off the hard work of the victims they steal from like slave owners striving to ensure victims aren't enriched. When the narcissistic thieves with mental illness are veterans with a predatory behavior great harm has been done to students' careers, finances, businesses, and personal relationships. A real threat to young students is older narcissists who are teaching at universities or running fraternities or sororities.

While the majority of professors and educators on university campuses are honorably professional, the real danger resides in the small number of narcissists. Narcissists have manipulated the hearts and minds of colleagues and Greek life students. Even a few narcissists on a campus can create major chaos upon young minds being deprived peace! Many educators have rejected academic professionalism, persistently stealing from students while being criminally narcissistic.

Narcissists who disregard the feelings of students are the driving force pressing victims into depression. The torment that narcissists inflict as cruelty upon victims has increased the number of suicides on university campuses. Narcissists stealing a student's intellectual property wealth has increased the number of violent incidents in our society.

In the narcissistic mind the game is winning dominance over another person's clout, wealth, merits, and works to validate themselves. Narcissists undermining students through deviously surreptitious behavior is dangerous in any educational setting. Students who plan to have children and spend they're time learning the proper ways to raise their offspring must be cautious of homosexual narcissists stealing their notes. Thieving homosexual narcissists have a demented angle to sabotage the way heterosexuals would raise their future children. Homosexuals undermining every thought that heterosexuals have is a real activity in our modern society. If you spent time in your life writing information to be a good parent beware of homosexuals stealing your written work. If you desire to get married beware of narcissists opposing this desire. If you desire to start having children beware of homosexuals surreptitiously planning to undermine your every way that you choose to educate your offspring by stealing your written work. Beware of homosexuals opposing your religious preferences and undermining your written plans in how you would raise your children in faith or academics.

Narcissists rejoice internally when harming students mentally, medically, physically, and even sexually on many universities. Some narcissistic teachers fulfill their unjust desires for financial gain by targeting brilliant, disadvantaged students, while striving to take advantage of them. Students need a protective community where they can study having productive socialization with other students. The consensus for institutions of higher learning must be that all students be given equal opportunities for participation.

The attitude of students in fraternities and sororities that they're better than certificate students creates division on university campuses.

Universities must not prevent certificate program students from establishing fraternities and sororities for they need socialization. There are some people who want to sabotage others from having socialization when it doesn't meet their dictatorial one way angle of thinking. A universities faculty that strives to block a student's socialization isn't healthy but very discriminating. A society's problems originate from its failings to not change personal interests, by refusing to grant permission to those students asking for new opportunities. Cruelty begets cruelty; so with so many suicides, shootings, and bombings in communities, universities need to change. Allowing students in certificate programs to purchase fraternity and sorority houses in investments to establish new social groups would be helpful towards improving our society.

Discrimination has no place on university campuses. Preventing international students' the opportunity for socialization isn't healthy. Students need fraternities and sororities to interact with others positively in their communities. How would you feel if you learnt a foreign language having a desire to belong to the group of people you attend school with, but were only treated as an outcast? Students don't desire to be separated, disenfranchised, and belittled on university campuses. Sometimes the very people who hold the highest honors academically are the most offensive in being cruel towards students striving to improve themselves. Students are no longer children; they'll behave as adults when harmed, insulted, or violated.

Teachers improperly using rearing techniques on students have brought us to this point in our nation where we witness too many violent events. Those who aren't licensed psychologists shouldn't be trying to use psychological techniques for they're inappropriately violating another person's medical status. Not respecting others' space and possessions has created violent problems in our society. Talented teachers with a real compassion for students will give each student space to create and share in their own time. Breaking a professional protocol structure for peace with students creates strife and hatred in our community.

Narcissistic teachers need to have their credentials of merit rescinded for any unprofessionalism towards students. When students commit suicide or retaliate in violence those instigating the trauma need to be held accountable. Lording over students is not a normal behavior for any educator. Students are to be given instruction in their learning and shouldn't be tormented by any person claiming to be a teacher. Narcissists renounce student achievements while they adore themselves with recognition and belittle victims they've criminally victimized with thefts.

I propose that program certificate students be permitted to wear caps and gowns to honor them. I also propose that certificate program students be permitted to create fraternities and sororities, which foster good accord among students living together.

In and outside universities, good student communication is crucial to each generation's humanity, living in a global society having unexpected discord and chaos. Certificate students belonging to fraternities and sororities will improve their international language skills beyond life at the university. Good student communication is very important for natural disasters, international business, technology events, and preventing wars that affect so many people. Preventing certificate students' an opportunity for socialization is inhumane. Universities need to change with the times or find themselves deprived of communication between students. This generation of young student leaders have a need to improve on their communication skills. Students that socialize more often are less likely to create atmospheres of conflict that can develop into massive storms with minds clashing into violence. Students today need social relationships outside phones, texts, internet and gaming platforms. When fraternities and sororities first established themselves long ago, these modern technological platforms were nonexistent.

I propose that certificate program students wear the traditional black gown but the tassel worn in ceremonies be divided in specific colors associated with their chosen course of study.

Take into consideration that certificate students often return to school to complete additional studies improving their education.

Graduation tassels divided into four quadrants will require four certificate programs to be completed for each color awarded in that course of study. If only one certificate program is completed, the tassel shall have only one color band joined to the remainder of the golden-yellow tassel or black tassel. For example a student studying screenwriting, directing, and film producing in the multi-media arts may decide to study for a (B.S.) bachelors of science degree in television to earn their four color tassel. The spectrum of light students need at universities for their learning opportunities must be multi-dimensional. Some certificate programs may be only a year in duration, while others can last up to three years.

The point is that if students are studying, they're becoming an active part of the academic community. Students should be honored, and not be treated as less important for actively contributing to a societies progression.

After completing their first course of study many students plan to return to school after getting some hands on experience. Students often want to try occupations to see if they like the work they've studied for, before they make a big long term commitment for their education. Each certificate progression warrants that additional honors must be awarded. In today's society, computer skills and coding are certificate programs with many different certification levels. Certificate students who complete many certificates should be permitted to wear multicolored cords and stoles with their graduation garments, in addition to their multicolored tassels.

Certificate students must complete all coursework to be allowed their ceremonial vestures. Rewarding good behavior is the right way to unite students with faculty, bringing a campus together socially, psychologically, spiritually, physically, and educationally. We need more positive school spirit on campuses and within our communities. With thousands of certificate courses being offered through universities, we need the student attendance in these programs.

Certificate students completing honors in applied sciences, academic research, and doctoral degrees must complete comprehensive examinations providing a thesis/dissertation paper for publication. Certificate students pursuing advanced studies shall be known as doctoral students or PhD doctorate students. Society at large needs to change being overly critical of the intelligence of others, for a student's mind often thinks differently than the way we think. Academic professionals with behavioral errors aren't contributing to the progression of nations. Many great innovators in history didn't possess a college degree and were active contributors to society, saving countless lives or changing technology.

As a student I first earned a GED and then a high school diploma. Then I took the initiative to take additional educational classes offered in the community for cinematography, as well as screenwriting having an interest in these occupations. I completed many different on-the-job training courses, gaining practical experience at a variety of jobs. Soon I enrolled in a college of technology for commercial diving, (ROV) remote operated vehicles, and underwater engineering earning a diploma. Then I worked on diving projects around the country mostly in construction and gold recovery. A paying contract opportunity arose in between jobs for scientific research so I participated in bio-astronautic training like a real astronaut.

Unfortunately I was injured in an automobile accident years later and learned how to walk again after my injuries healed. I couldn't work again as a commercial diver so in between jobs I spent much of my life's time writing for long hours completing drawings, sketches, and journaling details for filing patents. I compiled over thirty handwritten journals using my various talents in the arts, and they were all stolen. My cholesterol was elevated and my mind was disturbed as I struggled to piece together the remaining years of my life the best I could.

The road map we plan in life sometimes places us on a course we didn't intend to travel. When that life's journey is on a specific navigational route we must sometimes go with the currents flowing through the storm to reach our aspired destination safely. Finally I decided to pursue a certificate course at a university changing my career goals. There I discovered social exclusions being considered less important and devalued since I wasn't enrolled in a four-year study course.

I felt a condescending aura over my intelligence by the administration including the student body. I considered the fact that since I felt excluded then maybe others younger than myself also felt similarly excluded. Education shouldn't have social exclusions. Students need the opportunity to create social groups that foster a good foundation of acceptance. Not all social groups have the same methods of accepting members, which is fine, as long as other students can also have their own groups as well. I'm sure the film making students of the arts would like their own fraternity or sorority for creative minds like themselves. Creative minds have talented skills to raise the funds for their own dwellings. Why are some university campus administrations denying students a social existence to have a life, liberty, and the pursuit of happiness environment?

With all the division in religious beliefs, politics, cultures, and increased distrust amongst nations the students today need more socialization than ever before. We need a variety of changes to meet the needs of students who are socially deprived. Students excluded in high school don't want to find the same social exclusions at universities. We should honor and love students who make the conscious decision to make improvements in their education. Students desire to have equality and socialization is part of that equality. A domineering student body that makes certificate students feel excluded can be one form of hatred that's oppressive upon new students who feel the exclusion and separation.

Other students or faculty members making certificate students feel like they can't read, understand, or comprehend what their fellow students are studying is a belittling, and condescending.

The ideology that no one can do what another person has done is a super-ego delusion. Every year many students graduate with an assortment of university degrees. To be condescending towards others is an irrational thought process dwelling inside uneducated people acting like professionals. True savants can quickly detect those who aren't true savants. False savants wield a behavior that lacks qualities true savants possess. Wielding falsehood is something false savants project because they don't possess the genius mind others have and they resort to psychological projection as introjection. A true savant genius will have a mind that doesn't lack a compassionate understanding by omitting certain qualities others would with their thoughts. Many savants are analysts in their own way, which isn't defined by general definitions or classifications. True savants find themselves comfortable in the subject matter that makes them the most happy using their time according to their passion. Many certificate students, plan to pursue additional studies when finances improve and business practices become profitable. Some veterans and alumni already possess four-year degrees and are condescending towards new students planning improvements in their lives which is unprofessional. With all the violent shootings and acts of aggression happening in our society, we need to make students feel more accepted by providing them a safe environment. The right way to reduce violence in our society is by increasing socialization amongst people. Sociable people are less likely to develop anger because of mistreatment versus anti-social people who would encapsulate anger and then explode through violent acts. Some teachers have practiced extreme condescending behavior for a very long time toward's students and that's injustice. Military personnel minimizing a citizens' abilities to do they're job better is a great error in thinking because all military jobs can be done better. Billions of dollars of damaged government property can be attributed sailors and soldiers, not doing their jobs correctly. There are criminals in every military branch of service. Mentally disturbed soldiers have psychologically traumatized, assaulted, and murdered citizens. We need to uplift those contributing to society in any capacity, small or great, for that's what society consists of a diversity. Thefts by some Greek life students at universities have carried over into white-collar crimes outside university campuses.

The theft of academic creative works students have made has caused many problems in our communities. Thefts committed against students has increased homelessness, poverty, and depression in students forcing the victimized towards destructive lives by sabotaging their dream oriented goals. Expecting our society to conform to laws when American leaders refuse to obey an order of honorable conduct is illogical. First leaders need to obey the laws of a nation and not their own agendas which violate lives. The circumvention of truth is not a foundation to build a nation upon.

Lies build up a nation's anger, and anger leads to strife by contributing to mixed feelings which lead to violence, and that corruption has no productive place within our peaceful society. Universities shouldn't support liars or even endorse liars for meritorious recognition. America's foundations have lost integrity among the angelic overseers of this world.

Heaven has been witnessing the American murders, protests, wars, crimes, division, hatred, corruption, thefts, riots, and racism well-established in deadly sins. America's corruption has no place in the ecclesiastical realms where the holy commandments are obeyed. Any government, including the American government, that believes it can supersede heavenly laws and retain its power will quickly find itself uprooted. Like the floods which expose the roots of a tree causing it to topple, so shall nations be that deny holy eternal commandments given from Yahweh. No nation can be honored as royalty or have high status when it lets its own people live homeless, hungry, ill, or neglected. Jesus Christ attended to the needs of the poor in his holy ministry.

Many people within this nation have mental illness and believe they're royalty dwelling above others. The royal rank's among heaven's clouds reside with Saints and Angels. Repetitively sinning against the souls of others isn't a component of royalty. Heaven has a totally different ranking authority which cannot be compared to the ranking systems established on earth. Yahweh assigns ranks in heaven unlike individuals who assign rank here on Earth. All ranks come from the Creator for those who live holy lives, and there are some hearts who will never agree with Yahweh's decisions on how rank is distributed.

Many Americans cannot accept that they're living in denial. Americans continue to abuse alcohol and indulge in illegal narcotics or prescription drugs frequently. America's ongoing war is with heaven by refusing to live clean from addictive substances and abide by the holy commandments from the Torah. The angle demonic people have adopted in opposing heaven's souls is to steal from Yahweh's chosen people and to sinfully place themselves higher in a position than the saints and angels. We should all strive to be fruitfully righteous everyday for then our rank will be higher in heaven. Don't be like the wicked, sinking further into darkness where there isn't light. Drug dependency creates ingrained delusional thoughts in users which cannot be easily extracted or changed. Brains that have physically changed because of illegal narcotic use and alcoholism can never return one hundred percent to normality. The human brain scars when trauma is induced and often a certain degree of retardation envelopes the physiological processes within the brain. When mentally ill people huff or inhale aerosol sprays to get high brain cells are damaged. When drug users use bongs, pipes, joints, and other narcotic paraphernalia it's the same as huffing or inhalation of toxic fumes. Students who are victimized by narcotic users are having to contend with those who are mentally ill and should be in mental health psychiatric facilities.

When the American government refuses to hospitalize veterans, soldiers, and the civilians using narcotics they're intentionally putting citizens in harms way. The United States government should be prosecuted for failing to protect the American people from dangerous people who have destroyed the lives of civilians.

Often the United States government has chosen to ignore the laws it has made and just allow the criminals to continue victimizing others.

The American government doesn't want to settle repayment to victims for any cumulative punitive damages it's knowingly guilty for paying if it was to truly practice democracy. The United States Veterans Administration has hospitalized veterans who are psychotic. Rather than keeping veterans in psychiatric facilities under the healthcare of the Veterans Administration they've released veterans into communities to harm civilians lives. The United States government has concealed it's crimes against civilians while not supporting democracy in obeying the laws of this nation. When the victims are people aspiring to be on the list of the wealthiest people on earth then great is the damage of the stolen information taken by the veterans perpetrating the crimes. The United States government has improved it's infrastructure from the data it's acquired from thefts conducted by veterans divulging stolen data. The victims are all those people who were unpaid when the United States government made billions of dollars of improvements in security, and infrastructure from stolen data. Entrepreneurs victimized by thieves enabled by the United States government haven't been paid for all their brilliant intellectual property stolen and used by different departments of the US government. These aforementioned statements are the facts and the FBI hasn't investigated anything allowing the crimes against victimized civilians to continue. Veterans have digitally and analog transferred stolen data through clandestine methods to associates in the LGBTQIA community. Stolen data has transferred to family members of senators and other professionals who've been in on the crimes done upon entrepreneurs, innovators, authors, and designers. Those who created the intellectual property haven't been paid in this American democracy participating in the crimes against them through those they empower.

The world's military ranks are not recognized in heaven, for these transgress against the will of the Creator. People have been disobeying the six hundred thirteen inextinguishable holy commandments Yahweh's given humanity, and these commandments shall never be circumvented. Living sinfully, only perpetuates transgressions to adopt Satan's paths. For Satan chose to be above the holy saints, angels, and Yahweh in his evil transgression. America needs to support students praying the rosary like the holy children of Fatima, Portugal, seeking Jesus Christ's salvation.

Live like a saint every day so you'll be prepared for heavenly living.

Speak only fruitfulness from your tongue without deceit, for then heaven's ears will listen to your truth. Turn away from the ranks of this world for those ranks of the world have no ranking status in heaven above. Strive for heavenly eternal things because this world's inhabitants will steal the possessions of what you've created. Many in this world adorn themselves with riches while they cover their sins. Seek heavenly garments and holy heavenly crowns greater than those possessed by the fallen angels who dwell beneath Satan in they' re allegiance.

So many government employees, and military professionals with merited high ranking uniforms say they do all things well, while ships and aircraft full of supplies aren't sent to starving infants, children, and families in need. Jesus Christ would have sent supplies to the needy, so how are people in government claiming to be his followers? Soldiers have been quick to put down other people they think they're above, isolating equal opportunities away from many citizens who deserve to serve in the military. I expect to see more American soldiers being fought in the future due to the oppressive war tactics of suppression they've committed against citizens.

The American government hasn't shared military training, benefits, pay, and education with its citizens; but it has shared these things with foreigners. A system built on denying its citizens opportunities and excluding them is destined for failure. As time passes, no one wants one-sided groups of leaders hoarding all the wealth, education, and prosperity only for themselves or their kindred.

China has been involved in fewer international wars than the entire history of the United States. Yes China has wars with its neighbors but so has Europe, India, and many other countries. Those who land space rovers on the far side of the moon aren't entirely unintelligent. Maybe Americans need to look in the mirror and stop being so judgmental of others when they need to correct themselves. Nations around the world speaking different languages are viewing America as the mentally ill regime.

So many students could become pilots, astronauts, and space explorers, but racist people enforcing American laws for their own benefit deny opportunities for advancement to those seeking to have experiences also. Systematic racism in American hearts has enabled the denial of the people's requests and disenfranchised students. Meanwhile, government-employed Americans enrich their own relatives half the age of those minority students or citizens denied their requests. This aforementioned systematic hatred originating from the state and federal government will not stop and this is why America's destiny isn't forever.

Americans have been the pinnacle of oppression becoming like many countries around the world. Denying the dreams of others for multiple generations is what's going on in America. Americans will be remembered in history with a legacy of harming others that's carried before Yahweh's heavenly altar. Jesus Christ did all his preaching without causing harm to any one person. The American legacy has shown countless people being harmed while the holy commandments of Almighty Yahweh are not being obeyed. Coveting what others possess is a snare that reaps great misfortune. Going to war with those who have Yahweh's support with heavenly weapons possessing more power than you have is fruitless so why are Americans challenging the power heaven can wield? Americans cannot continue sinning without the time of mercy expiring on those who've hidden their agenda to fulfill sinful desires. Americans who want to dictate how the Holy Spirit works in the lives of others are serving Satan; for the souls of others aren't controlled by sinners with minds bound to evil intentions.

Nations pursuing peaceful space advancement is a good idea. Let's have a global competition in which students from every nation can submit outer space architectural designs at a conference attended by United Nations General Assemblies. Let's get young students designing new buildings and cities, for Mars being constructive rather than destructive. Let's view the student submitted competition designs at the United Nations by coming together rather than apart. Student project endeavors can be unveiled at the United Nations for outer space, Mars, lunar colonies, and the moons of Jupiter. We can have one to four categories for the United Nations competition.

Let's put effort back into peaceful student competitions and education, so many nations can emphasize productivity for space exploration. All these student models will be beneficial for science, technology, engineering, and mathematics to encourage new development opportunities. Third-world countries are slowly getting out of past ideologies, knowing the importance of education for all. The superpower nations should unite in helping third-world nations with medicine, housing, sanitation, and agriculture programs coupled with education. Design parameters will bring forth a new generation of space inventors because these competitions will help Earth-based improvements in many nations. Let's discover which nations have the smartest students. We need to discover great political leaders and smart innovators for advancement. Through space projects let's see students contributing to the benefit of mankind here on Earth and beyond.

In each nation, there will be a competition among students for the best space designs. The top one to four designs get sent to the United Nations for world leaders to view, and the public can judge online, viewing the concepts in real time. Since world leaders will already be attending the United Nations General Assembly the transportation costs are covered so they can just bring along the models. The models will first be X-rayed and then be covered by a cloth moving them to the display room up to the point of the unveiling.

World leaders can hype up their students' designs until the model unveiling. In the great hall world leaders can create an aura of competitiveness in the political spectrum. The creativity of each nation's elite students showcasing their models will show leaders how important it is to go to outer space. Billionaires who've been encouraging space exploration can attend the United Nations summit.

Some billionaires would be willing to offer secret prizes that are hidden from world leaders for the best models which incorporated smart technology in the design. The hiding of the names of students and the countries the model originated from is part of the mystery. In the completion the concealing of data who created the model will help prevent any bias judging or favoritism from taking place.

Prizes can be tax deductible since the United Nations is affiliated with non profit charity organizations. Also prizes can be scholarships, internships, or even monetary purses to help students' lives. The top ten designs from the competition will be awarded prizes. World leaders can judge who has the best model for they're vote. The general public can judge who has the best model for they're vote. The two separate votes will then be analyzed in a comparison column over which models were most favored. The political vote is that of the world leaders and the popular vote is that of the general public. No one but the billionaires will know if the prizes are small or large till the awards are presented keeping the competition mysterious as well as exciting.

World leaders from each country can walk through the rows of unlabeled model designs and be enlightened by the vast diversity of creations students have made. Students are thinking about living off the earth. World leaders must embrace humanity's future of living off the earth in their students' dreams. Establishing space colonies on other planets is the exploration we need. These fun forms of competition are a great way to establish peace among nations.

This competition will be a shared experience for many countries. When world leaders at the United Nations can start respecting each other, this is progress toward world peace!

Uniting the world on common goals like space exploration, and new world habitation through creativeness is what all humanity needs. Nations need a new avenue for the advancement of humanity, and space can be that avenue. Why build up a world with new ideas then destroy the world through wars? War is nonsense behavior lacking true intelligence to adapt to problems with better more viable solutions. Stop striving to build up nations for war, and start striving to build up nations for living in space. The establishment of new colonies for humanity's survival beyond this world's destruction is a smart contingency protocol. Earth was hit by asteroids and meteorites in the past so it will happen again. A nuclear war will put humanity and so many species into extinction and this is foolishness. Space exploration aims our combined energy towards constructive development processes and not towards extinction.

All Christians, Catholics, and Jews believing in Jesus Christ and who obey the heavenly holy commandments are one body. I hope many parts of this world can be saved before some government people destroy all the miraculousness beauty of the Earth. Governments are too much at odds with one another. The focus needs to be taken off each other politically and we rather need to put our energy into outer space engineering.

In our modern society, the educational system is flawed because students can act out without being punished. How about having teachers in classrooms wearing body cameras, so when students misbehave, a quick push of a button, and the video is sent to the principal's office? Parents can watch the videos, and their child will not have an avenue to circumvent being punished.

Misbehaving students are indicators that something else is going wrong at home or within the school. Often the problem is overbearing parents who want to be over controlling thus messing up their child's thoughts with improper guidance. With the introduction of teacher body cameras, inner-city school systems can have methods to thwart bad behavior. School systems have seen gang problems, illegal drugs, smoking, alcohol use, sexual advances, and disruptive students. Video monitoring systems in all classrooms help increase student safety through prevention, and minimize abuses from students or from teachers.

The federal government should fund the teacher body camera system for student safety because of all the school shootings. Psychologists can review videos intermittently and begin the process of discovering which students have a propensity for violence. Discovering mental illness requires a team effort to uncover the indications of deeper psychological problems. Billions of dollars are spent on defense budgets while thousands of young people have died from gun violence. Mass shootings should be covered in part of the national defense budget. Stopping terror acts at schools K-12, universities, and in our communities means we need better character improvements. To thwart disturbed minds from generating hatred that leads to violence we need social character development programs.

We need to honor our productive students and we must also protect them from harm. It's not the guns creating problems; it's the people behind the guns who are creating problems. If the government starts taking away guns, then mental cases will want to murder others by other means.

We've already seen violent people running over strangers with vehicles and creating bombs to explode upon civilians. How do we stop students or people in general from wanting to harm others? We stop violence and anger by making people happy so that they no longer want to harm others. We need to put emphasis on programs that make young people and especially adults feel more happy as something to belong to. We want people feeling good because they're not going to go murder and create chaos in our society. We don't want to make sociopaths or psychopaths to feel good because that can be extremely harmful since they get pleasure in disturbing others and inducing harm upon others.

High schools and universities need more social opportunities with extracurricular activities. Taking away a person's ability to find a conduit for happiness is inadvertently creating a space where a person can feel hated, rejected, and isolated. Why is there so much violence? The answer is because the perpetrators aren't happy. How do we fix the problems in our society creating discourse? We create the necessary community facilitation to enable sad people to have socialization for happy friendships inviting them to participate in recreational activities. Never take away what makes others happy without consequences; that's basic childhood rearing not psychology.

Not all teens and adults react the same way to disturbances, and this can create unwanted turbulence. If a person's intent is to pick a fight, they shouldn't be astonished if the person they're disturbing doesn't fight back the way they wanted. There's too many people with mental illness constantly picking psychological fights, and victims disturbed can sometimes snap to commit violence. Even the Holy Bible says to not stir up strife, but there are some people with mental issues who constantly want to stir up strife. Our educational systems need to recognize the different types of personalities mentally ill residents can harbor, and this is to quickly avert unwanted violence in our communities.

What one person thinks is a sufficient amount of violence to render balance, a totally different person may think twice or three times as much is needed to resolve the same situation. Your scale is not my scale, and my scale of judgment isn't your scale. This is why it's very important to not mess around with people's feelings, emotions, sexuality, religious beliefs, and psychological well-being. What you may think is funny may also be what sends the next person down the road to committing a mass murder. All students have different forms of suffering in their lives, and some students have had more suffering than others.

Be careful which paths you tread on life because what you do has repercussions in our community. The world in which we live is connected, and your actions can quickly be communicated around the world to others. If you find enjoyment bullying others being the type of person who violates another student's personal space then please don't try to attend a university.

Our public education systems and universities are for students seeking peace within their community. Strange people who want to torment the mind of others psychologically and disrupt the learning environment aren't welcome at universities. Many disturbances from strange people have destroyed student ambition sending them down a path for violence, they wouldn't have taken if they weren't being violated. America needs to learn from the past and start making changes for the future. It's nice for students to belong to a educational program which enables the building of stronger relationships fostering more productive happiness on our campuses.

Life, liberty, and the pursuit of happiness—so why, then, are there Americans telling certificate students they cannot establish fraternities or sororities? The rules of the past have led us to the point where Americans hate each other. Americans are willing to protest or even clash with police over their rights being violated. We have the answers to fix problems; and now we need to implement changes nationwide. If America doesn't change, it will repeat the heartache conjured in its past, instigating violent episodes to recur. I believe the main problem in America is the implementation of hard work on good ideas.

3

MUSEUM OF THE HEALTHY AND DISEASED

Drug and alcohol dependency is a choice that could be avoided by billions of people worldwide. Establishing a new museum within the nation's capital focusing on health, numerous diseases, and drug prevention is mandatory move toward progress. In every city across the United States, we all know a drug user, alcoholic, or person living an unhealthy lifestyle requiring an intervention. In America, we see homeless encampments with drug addicts with various diseases, and that's a situation requiring an intervention.

Growing teenagers in every city must continue to go to school, learning facts on health and taking field trips to advance their education. The sad reality is whether it's children, mothers, fathers, siblings, acquaintances, neighbors, or relatives, we're all exposed to addicts making unhealthy choices. One of the most straightforward methods of teaching young teenagers the risks of disease is by exposing them to exhibits that provide comparisons. In one jar, we have a healthy, normal specimen, and in the adjacent jar, we have a very badly diseased specimen. When we show students how different diseases damage tissues and organs of the body, we educate them for their future endeavors within our communities.

This museum shows a healthy organ from the body preserved in a glass container or jar, and immediately beside the healthy organ is a series of diseased organs in jars or containers. Beside each container, there will be printed medical facts about each anatomical part, including descriptive physiology. The museum will be filled with hundreds of specimen containers, including dissected organs and tissue slices, giving visitors an interactive learning experience.

Organs affected by disease will be placed directly beside healthy organs, visually depicting contrast under magnification lenses. We want students to learn what healthy tissue looks like and what diseased tissue looks like. We want visitors to learn about the human body: its entire anatomy and the pathology affecting tissue's physiology. We want student visitors to start thinking not just about the tissues but also about the behavior of diseases. An illness's early stages of development in turn effect a tissue's progression into becoming an uncontrollable disease saturating healthy cells with toxicity.

Human beings and zoological animals can pass diseases back or forth among themselves. I believe it's important that the museum also have a section dedicated to animals with various diseases that can be transmitted to humans. Showing a perfectly healthy animal organ directly beside a diseased animal organ is also educational. Our focus should be on preventing pandemics in the future. With pandemics affecting a global population and families living together in confined spaces, we need changes to our health care. Insect species can spread diseases to humans, and these diseased organs or tissues can also be displayed in the exhibit.

The purpose of the museum is to educate school-aged children and adults. We'd like to provide a deterrent showing how drugs, alcohol, smoking, vaping, and other unhealthy lifestyles can be harmful to human tissue, organs, bones, and blood. There are so many legal drugs and illegal drugs and types of alcohol, each affecting organs differently. City-run morgues, hospitals, and emergency services deal with overdose calls regularly, and these body parts can be obtained for scientific dissection for our museum exhibit.

The museum can generate money from ticket sales. A portion of ticket sales can fund different health-care projects, which can include surgeons performing emergency surgeries. Every year, schools from many cities nationwide and internationally can visit the museum through student field trips. The medical exhibits will be open to the public so visitors can learn about diseases. Educating children with visual contrasts depicting healthy tissue and unhealthy tissue is a simple method to bring awareness in an impactful way. Optional would be the federal government creating the Museum of the Healthy and Diseased for the National Mall. This museum, of course, would be a new, weird exhibit with strange specimens.

There are so many tissue-destroying bacteria, viruses, fungi, germs, and organisms that can cause many different diseases listed in an encyclopedia of infectious diseases. Our museum needs to explore all the different diseases of our world by having specimens of all of them for study. We need to start the process of finding cures for every disease known to humanity. The focus must be on zoological animals spreading infectious diseases to humanity. We also need to start finding cures for animal diseases because that may be the missing link to finding cures for many of the diseases human beings now have. We're interconnected with our world, including the various species within it. Our exposure to numerous species throughout the planet in many nations and territories means we need to trace more epidemiological origins of pathology in the laboratory. Tracing disease origins can begin from interest started at a museum.

Illegal drug dealers have been mixing together so many different drugs and testing their concoctions on users. This doesn't just cause overdoses; it also causes new diseases to emerge. Diseases may find certain compounds in mixed drugs advantageous for them to feed off, providing nourishment for metastasizing. Chemical compounds in different drugs can initiate the growth of a disease or aid its spread to other parts of the human body. If a person survives illegal drug use, often the consequence is permanent brain damage, tissue damage, or other physiological changes.

Once the human body's nerves are damaged or altered by drug use, they may experience seizures, spasms, pain, sensation loss, and motor skill function disorders. Why aren't antibiotics as effective as they used to be? The answer is that drug dealers are tampering with prescription drugs by exposing drug users to an assortment of chemicals that diseases familiarize themselves with. Elderly patients have their prescriptions stolen by handymen, landscapers, and church helpers.

Illegal drugs and prescription overdoses from experimental drug interaction combinations include: marijuana, hashish, heroin, opium, cocaine, amphetamine, methamphetamine, methylenedioxy-methylamphetamine (MDMA), Flunitrazepam, GHB, ketamine, PCP, *Salvia divinorum*, dextromethorphan (DXM), LSD, mescaline, psilocybin, anabolic steroids, inhalants, Actiq, Adderall, alprazolam, Ambien, amobarbital, Amytal, Anexsia, Antabuse, Ativan, Avinza, Biocodone, buprenorphine, butalbital, butorphanol, Campral, carisoprodol, chlordiazepoxide, clonazepam, clonidine, codeine, Concerta, Damason-P, Darvocet, Darvon, Demerol, Depade, acamprosate, Desoxyn, Revia, Dexedrine, dextroamphetamine, Dextrostat, Di-Gesic, diazepam, dicodid, Dilaudid, disulfiram, Duodin, Duragesic, Duramorph, fentanyl, Fioricet, Fiorinal, Halcion, Hycodan, Hydrococet, hydrocodone, hydromorphone, Kadian, Kapanol, Klonopin, laudanum, levacetylmethadol (LAAM), Librium, lorazepam, Lorcet, Lortab, Luminal, MS Contin, MSir, meperidine, methadone, methadrine, methaqualone, methylphenidate, morphine, naltrexone, Nembutal, Norco, Oramorph, Orlaam, OxyContin, oxycodone, Palladone, Panacet, pentobarbital, Percocet, peyote, phenobarbital, Quaalude, Ritalin, Rohypnol, Roxanol, Roxicodone, Ryzolt, secobarbital, Seconal, Soma, Stilnox, Sublimaze, Suboxone, Subutex, SymTan, Temesta, tramadol, Tramal, triazolam, Tussionex, Tylox, Ultram, Valium, Vicodin, Vicoprofen, Vivitrol, Xanax, Xodol, zolpidem, and Zydone.

Minimally intelligent corrupt drug dealers combine drugs to create new chemistry chains or mixtures and then test their experimental concoctions on addicts. Often, the addicts are deceived or forced into taking these combinations as payment for other drugs consumed. These aforementioned illegal drug activities are not going to stop. Bringing awareness to students early on is the only way to combat the criminality of illegal drugs doing harm on our society with addictions.

Illegal drugs have been a contributor to murdering millions of people and ruining countless lives. Illegal drugs have torn families apart. While morgues fill up with overdoses, drug dealers frequent nightclubs, living the good life with fancy cars and plenty of cash to impress the ladies. Often pretty ladies are swayed into corrupt adult entertainment industries because they see drug dealers with wealth. Criminals getting victims addicted to drugs then manipulating weak minds into the adult film industry is a massive American problem. The problem with illegal drugs has led to sexually transmitted diseases being spread around the world more intensely among international nations. As riches increase among partakers in the illegal drug industry, they've chosen to travel outside the United States to poorer countries with their sexually transmitted diseases.

America's porn industry international travelers are infecting children, teenagers, and adults with STDs in multiple countries. American drug dealers have preyed upon nations with children, teenagers, and vulnerable adults living in poverty for over a century. Foreign countries have been plagued by massive numbers of their citizens being infected by Americans with HIV as well as many more sexually transmitted diseases.

Statistical information needs to be established for the medical community to reveal the accurate data regarding what races have spread sexually transmitted diseases to other ethnicities, thereby contributing to deaths, infections, and exposures. Medical statistics must be complied to diagnose where problems are originating from so medical professionals can address and correct these issues with accurate data. When medical data only shows us the truth, then there's no room for racism, discrimination, or prejudicial behavior toward any ethnicity. Medical doctors must record the facts of who has infected others to compile data to protect others from infections that can harm human life. Spreading diseases is attempted murder upon the victims who are infected.

Americans traveling to foreign countries and spreading infections has led to children born with harmfully rare conditions and deformities. Sexually transmitted diseases combined with illegal drug use are very harmful to the normal development cycle of a zygote. Dangerous Americans are doing drugs, going to foreign nations with infections, and impregnating women, exposing them to assorted chemicals. These chemical exposures affect the metabolism and tissues during fetal growth, which can cause infants to be born with medical conditions, developmental problems, psychological issues, or psychiatric conditions.

I felt it important to vocalize this point because in foreign nations there's a problem with serial killers, drug dealers, and mastermind criminals with a heredity that can be traced back to many nations, including America. Changing the future of all the world's inhabitants means enacting change now and not tomorrow. All civilizations need to put in their fair share of hard work to make improvements by caring for one another. A global standard for health care must be upheld by the United Nations and adopted by all countries to protect their citizens.

Parents need to better educate their teenagers about sexually transmitted diseases, including the infections they cause. A preservation museum can display healthy body parts in photos, videos, and containerized chemicals. The museum of the healthy and diseased is a new educational attraction that needs to be made.

Healthy organs can be placed inside containers directly beside diseased anatomical parts or bodies with sexually transmitted diseases. The purpose of this museum is to create a deterrent for youths wanting sex too early, not fully understanding the dangers associated with sexual intercourse in intimate relations.

The intention is to bring a strong medical awareness in health class to abstain from sex, thus reducing American infections from STDs. Yes, this display will be graphic and considered gross anatomy, but these diseases are real. Youths need to become visually aware of these dangerous diseases, and seeing diseases up close is a wake-up call to live more responsibly.

Teaching prevention to eradicate ignorance is very important in saving lives, as is protecting innocence from harmful sexually transmitted diseases. Visits to these types of museums will help aid humanity in the reduction of diseases spreading worldwide. Youths will learn the importance of practicing proper hygiene, improving the sanitary conditions for toilets, restrooms, showers, and touchable surfaces, which often transmit communicable diseases.

As humanity becomes more aware of the risks of infectious diseases, staying clean helps reduce potential transmission carriers. We must help physicians who need an accurate public communication system alerting them to potential hazards and onset exposures. People who can recognize that something is wrong are much more effective in helping resolve problems in their communities through effective communication than by ignoring the problems altogether.

Students and adults visually learning the dangers of dormant diseases will become more aware to risks. The more cautious one becomes with infectious diseases in one's life, the better one's life choices. As lifestyles and activities divert away from unhealthy decisions, our populations will become more healthy, enhancing the overall health-care system. A world with better health care for all is a path toward world peace, thereby establishing common ground with all nations.

Visitors to the museum exhibit can learn about viral, bacterial, and parasitic diseases. Virology and pathology will be introduced with many interactive displays throughout the museum. Toxicology will be introduced, showing necrosis of cells and even the disease itself as death or gaseous gangrene attacking healthy tissues. Videos of various diseases multiplying exponentially will help teach teens the dangers of being careless with their sexuality and choices.

This science museum will have many interactive exhibits that can be entertaining. Exhibits will display photos, specimen jars, petri dishes, pathological microscopy images, many infectious videos, and infected persons speaking prerecorded messages. Multiple-choice questions on digital-display screens shall accompany many exhibit sectionals.

Any newer technologies to educate visually, audibly, and interactively shall be used in the museum displays. Sexually transmitted diseases observed in the museum will include but not be limited to trichomoniasis, urinary tract infection, syphilis, scabies, pubic lice, pelvic inflammatory disease, molluscum, lymphogranuloma, herpes, hepatitis, granuloma, genital warts, gonorrhea, chlamydia, chancroid, bacterial vaginosis, Acquired Immune Deficiency Syndrome (AIDS), vaginitis, yeast infections, and human papillomavirus.

The museum will have preserved human tissue samples, fetuses, organs, body parts, disease samples, and encapsulated diseased cadavers for up-close observation, much like a Body World exhibit, but in embalming fluids or specialty solutions. Some tissues or cadavers can be chilled for display or preserved through new technologies, preserving virulent qualities for visual observation. Students may be asked to write a paper on their experiences at the exhibit.

A mixture of glutaraldehyde, formaldehyde, formalin, methanol, solvents, and a variety of other chemical solutions may be used in the exhibit preservations. Chemical injections and oil plastic compounds may be used for the preservation of necrotic diseases so that specimens can be visually analyzed in the exhibit. These exhibits of diseased anatomical parts in encapsulated chemical compounds are for research purposes, preserving their necrotic state for close-up observation of their structures.

Video recordings of patients with various mental illnesses can be displayed on television screens so students can see and read about a variety of diagnosed cases. There can be diseased adult brain slices preserved for students to see the major differences between a healthy brain and a diseased one. Visitors can interactively touch multimedia computer screens, learning from videos, diagrams, photos, containment preservation displays, and information written over imagery regarding diseases in the exhibit.

Potential mental conditions that can be examined in the exhibit include schizophrenia, psychosis, delusion, mood disorder, autism, bipolar disorder, borderline personality disorder, depression, catatonia, delirium, fugue state, cognitive disorder, conversion disorder, Cotard delusion, panic disorder, post-traumatic stress disorder, attention deficit hyperactivity disorder, schizoaffective disorder, major depressive disorder, development disorder, psychogenic amnesia, impulse control disorder, paranoid schizophrenia, depersonalization disorder, psychopathy, autistic spectrum disorder, pervasive development disorder, generalized anxiety disorder, insomnia, stress, confusion, stereotypic movement disorder, obsessive compulsive disorder, intermittent explosive disorder, fatigue, narcissistic personality disorder, and mental disorders diagnosed in childhood.

The preservation museum must contain a medical vault of diseased human tissue samples. The medical preservation vault must include biopsies, cell cultures, amputated limbs, dissections, embalmed body parts, and anatomically infected tissues from every infectious disease in our world. The museum's goal is to collect diseased human tissue with the worst appearance in gross anatomy. The human tissue and specimens that are the most severely diseased will be preserved in formaldehyde or embalming fluid for educational purposes.

Students and adults will be able to view diseases up close, with curiosity hopefully initiating a desire to find a cure. Macrophotography and video will be used to bring images of diseased tissues closer to students observing specimens. The museum will provide medical doctors via web videoconferencing and webinars to explain their study of various infectious diseases.

Students will be able to interact with the medical doctors through videoconferencing, which will be integrated into the exhibition hall's displays. Illnesses will be explained by medical doctors in videos, presentations, photographs, slideshows, microscopic magnifications, and exhibit preservations holding diseased body parts in fluids. There are many different types of medical doctors, and each one shall explain what they do in their labs or hospitals to deal with infectious diseases.

Some doctors will prerecord videos to teach throughout the exhibit in different sections. The different types of doctors will be explained in detail throughout the exhibits, for each medical specialty is different. Diseased flesh from patients who have passed away and donated their bodies to science will reside directly beside healthy body parts or corpses for up-close comparison.

Students and adults doing examinations will be able to look through microscopes at each virus, microbe, germ, bacteria, or organism creating disease beside each preservation container. Seeing infectious diseases alive or dead through microscopes is a very effective way to inspire continued learning in students. Infectious contaminations passed on by food, water, surfaces, insects, zoological creatures, and human transmission will be explored as well.

Students will learn about medicine in a more unrestricted manner, with teachers helping guide students throughout the museum. As the students start analyzing infections, some may have an aptitude, gravitating further toward a deeper interest in infectious diseases.

My mother passed away from cancer when I was a teenager because her life was stressful, and she smoked cigarettes. We thought the surgery wasn't going to be a problem, but we were wrong. She developed complications, which worsened until I lost my mother. This is another aspect that the museum can put emphasis on, asking students questions about what they would do if complications arose during their surgeries. The doctors who are available via web conferencing will be able to explain what they do when complications arise during surgeries. This statement is important because then students will begin to think about how they can improve medicine, including surgeries.

Health care is such a vast field with so many different occupations that not everyone thinks the same way. New procedures, inventions, and pharmaceuticals come forth frequently, changing medicine. The sooner we engage students with a desire to learn medicine, the sooner we can expect to see medicine improving. Fresh new minds will think differently than well-seasoned minds, and an approach to doing a medical procedure may change entirely with a new outlook.

Some occupations on the medical front will be discussed such as surgeons, anesthesiologists, gastroenterologists, nephrologists, urologists, rheumatologists, oncologists, neurologists, psychiatrists, ophthalmologists, cardiologists, endocrinologists, dermatologists, pediatricians, microbiologists, hematologists, obstetricians, gynecologists, pathologists, virologists, epidemiologists, immunologists, allergists, otolaryngologists, pulmonologists, and internists. Physicians providing their experiences is important for students.

A section of the museum can have information on genetics and neuroimaging, providing photographic comparisons of many diseases harming the body. Epilepsy, Parkinson's disease, diabetes, and multiple sclerosis can be examined, along with other diseases in the museum exhibit.

Diseases from around the world that are often not spoken of but are in medical encyclopedias with photos and statistical information can be studied in the museum exhibit. Diseases from many countries can viewed visually in the medical containment preservation encyclopedia of learning. As more illnesses emerge in our communities, we must protect the public by inhibiting the spread of infectious diseases.

This museum can also educate young people about how communicable diseases spread. This museum can educate on how to properly disinfect surfaces to prevent the spread of communicable diseases. Sexually transmitted diseases must be mitigated because humans can be infected with more than one pathogen simultaneously.

More than one virus infecting the same host can increase the probability for a genetic mutation to occur, making viruses slightly different than other strains of the same category. Many herbivores have developed a natural resistance to fungi infections over thousands of years, but when a virus is introduced, they adapt metabolically. The way animals metabolize energy is different from the way humans do.

To ingest any animal species that possesses viruses is dangerous and can lead to an infection. When we harm our environment with chemicals, we're also contributing to what an animal, insect, or other organism may ingest, thus changing their metabolic processes. Creatures dying because of environmental contamination is catastrophic to the ecosystems of the animal kingdom. Creatures that have adapted to environmental contamination by repetitive exposure can become catalysts that can possibly infect humanity with new viruses.

A virus exposed to a variety of environmental chemicals to which a host creature is also exposed may adapt or change over time, becoming a new strain. As we harm the environment around ourselves, don't be surprised that we may also be harming ourselves in the future with new viruses that emerge, infecting global humanity. The coronavirus COVID-19 is only one virus, and we still have billions of microbial human pathogens on Earth that have the possibility of mutation.

Diseases from insects that harm humans can also be displayed in the exhibit with images or encapsulated specimens of the insects. Information and videos on various insects that spread diseases, like mosquitos, fleas, or ticks, can also be included in the exhibit. Displays showing mosquitos—where they breed and what they carry—can be in the exhibit with statistics on the numbers of people harmed by these insects. Mosquitos carrying the West Nile virus, Eastern equine encephalitis, malaria, dengue, yellow fever, and other diseases can be displayed.

Human body parts and fetuses can be displayed, showing the harm caused by insects exposing humanity to diseases. Statistical information on disease deaths worldwide from various sources can also be displayed, with annually changing numbers in comparisons. The museum can have prerecorded videos from people who are suffering from various diseases, giving their heartfelt short stories telling how they contracted their disease or diseases.

These heart-wrenching stories will be the educational prevention needed for young people to reevaluate how they live their lives. A day trip to this museum will change lives. As visitors view the worst diseases in the world captured through macrophotography and microscopic video display, each person will begin to develop a new perspective on medicine. Factual medical information is so very important for students learning about medicine, and keeping that information as accurate as possible is a prerequisite for success.

We should stop people from spreading false information when it comes to the health of people in our communities. The home remedies people concoct aren't going to protect them in a pandemic when they need to be listening to physicians who are experts in their fields. It's our duty to educate people with facts, especially when it comes to their health and the health of others around them.

The internet has provided people so many different versions of how the pandemic originated and how the virus is contracted. Especially now, humanity needs facts when it comes to pandemics to protect families in our communities from harm. People haven't been considering that maybe the pandemic virus could pass to insects that come in contact with human blood.

If the virus does enter insects, will the virus adapt inside its host to become slightly different from its original version? A virus that has a bunch of spiky fingers may decide to change its fingers to become wider or stronger. A virus that mutates and unleashes a new harpoon-like structure with barbs protruding into cells in a new way could be devastating.

We aren't just dealing with the virus itself but the fact that individuals as well as governments may look at the virus as a potentially new bioweapon for use later as it's modified. It's important for students to see in the museum infected insects and how they behave. We must educate students because then they can become the future inventors, scientists, and physicians coming up with new ways to treat patients.

Encapsulating live insect specimens and quarantining them in a container with macro video, then showing students the insect activities up close, is important. The more we know about the animal kingdom and insect transmissions, the better we can prepare to protect ourselves. Manipulating the anatomical parts of the insects can be part of the exhibit. Students can see diseased flesh alongside insect cases giving statistical information showing how millions of human deaths are caused by the insects. The growth of organs can be explained in the museum for the purpose of inspiring the longevity of human life. The US Congress can bring together physicians along with myself to design and create this museum with education and National Institutes of Health funds.

Should the US government have no interest in funding a new museum and creating construction jobs in Washington, DC, then I would be open to private investors assisting me on this project. Students could easily spend three days in this museum learning about so many medical occupations, good health, and the dangers of infectious diseases. The National Mall in Washington, DC, is an international tourist attraction as well as a domestic attraction for millions of Americans.

To have a museum filled with gross anatomical parts of diseased human flesh would be great new attraction for students nationwide as well as internationally. The museum should have multiple floors to provide room for new or changing exhibits. The museum should have event space for physicians, scientists, inventors, professors, pharmaceutical companies, and private groups to use for elegant parties or conferences. There should also be a theater that can be used by lecturing professionals and students who need instruction or educational films on medicine.

The museum should allow for social events in the main exhibition halls and throughout all the different sections. If there needs to be a meeting with the media regarding health-care concerns, the theater space can be used to host question-and-answer interviews regarding infectious diseases. Live news can be broadcast from the museum theater during interviews, web conferences, university student medical shows, and events. The museum will be available for use by the Centers for Disease Control and Prevention, the Department of Education, the National Institutes of Health, the National Science Foundation, universities, and colleges. The investment in building a new museum should allow versatile use, enabling the space to be used by many groups, both inside and outside the nation's capital.

There should be park land set aside around the museum that can easily be converted to allow for future expansion of the museum as medicine changes. The historic ways in which we dealt with diseases were much different from the ways we deal with them now and will in the future. The museum will accommodate information about historical scientists, physicians, and laboratory technicians who contributed to infectious disease research or cures. Vaccination history and those who contributed to breakthroughs in medicine should be exhibits in the museum as well.

The actual diseased body parts viewed in the exhibit would be behind thick, unbreakable glass. Students can use viewing goggles to see inside some diseased specimens preserved in the containment cases or held in fluid suspension in glass jars. Fetuses and infants who are diseased can be also part of the gross anatomy reproduction section of the exhibit. Many diseased body parts will be very graphic, and some students may opt out of those sections of the museum if they feel uncomfortable viewing gross anatomy. Signs on the walls outside certain sections of the museum with graphic or gross anatomy can provide warnings to visitors who may not want to see such grotesque diseases. These warning signs outside certain sections with grotesque anatomy will help teachers, chaperones, and guides giving tours to students or visitors.

There should be a museum section dedicated to each medical occupation and branch of science dealing with infectious diseases. Some visual displays for learning can provide very close-up views of some infectious diseases engulfing living tissues in petri dishes so students can see how viruses or bacteria behave. Students can look through microscopes on exhibit displays and have a real-time digital connection to another microscope being used by an actual scientist or physician in a laboratory. Also, the microscopes could have an audio feed to the real scientist or physician wearing a headset and viewing the viruses or bacteria in real time. Students can communicate with the scientist or physician in real time, and the professionals can explain to the students what they're seeing happen at that very moment. This type of interactive learning display can be integrated throughout certain sections of the Museum of the Healthy and Diseased.

The more interaction we give students with real professionals, the more students can be inspired by the professionals, so students don't choose the path of illegal drugs or alcohol. Internationally, all teachers have a responsibility to keep their students off illegal drugs and alcohol. While we are trying to do the right thing to educate students in a positive way, the criminals are devising ways to harm students. It's extremely important that we make this museum available to every class of students nationally, along with all the other National Mall museums.

Government-sponsored field trips maybe the only vacation experience some teenagers have in America, and there should be funding set aside for these field trips. We desire young people to learn to become contributors to our society for our nation's patients, families, and the elderly in each generation. When we start young students down the right path early in life, this is the pinnacle period for them to carry these values throughout the remainder of their lives.

The numbers of infectious diseases will be updated yearly for accuracy in the museum exhibit. In a collective series, doctors in lab coats from various medical professions can explain what they do briefly in videos so students can learn about their occupations. Also, the growth of organs and healthy tissue can be explained or displayed in the exhibit as an introduction to how we can combat infectious diseases in surgery. Researchers who study epidemiology, pathology, toxicology, hematology, and virology, for example, can explain in videos what they do in their laboratories. Sometimes, students need to be shocked with images of real-life events, such as morgues or mass graves full of diseased human bodies. When the gravity of death isn't minimized, then the importance of life can be maximized.

Exposing students to these instructional videos explaining how many years physicians studied to become doctors will encourage students to continue their education. Showing the inside of psychiatric facilities and what psychiatrists do in instructional videos can help students become better Americans. Students can learn about psychiatric conditions to become better citizens socially. Students who learn about mental illnesses can become better at recognizing mental illness in their communities.

Students are already exposed to homeless people in their communities who use illegal drugs and alcohol. When we teach students about mental illness, they can, in turn, make better judgments to defend themselves from bad situations in which they could encounter mental illness in their communities. Students these days are exposed to mass shootings, suicides, pandemics, homelessness, bullying, and dysfunctional families. Improving the behavior of students is important.

We improve student behavior through teaching videos about the various types of mental illness so if they recognize this in themselves or in others, change will begin. Students with mental illness don't want other students to see them as having mental illness, so they'll make changes to conceal certain behaviors. The concealing of certain mental behaviors may be exactly what teachers need to control students who act out in class or disrupt other students. Students recognizing a mental illness may save their lives and the lives of others around themselves by preventing a deranged student from murdering them. Typically, teachers of the past wouldn't teach students about mental illnesses, but I'm an advocate for change. Mental illness is so rampant in our communities that we need to teach students about the dangers of people who have these illnesses. Health class needs to incorporate more lessons on various different forms of mental illness.

Governments have said to shelter in place and to self-quarantine during a pandemic. During a pandemic, the rich have homes, but renters and the homeless are left to scavenge whatever resources they can during a pandemic. Not having homes puts more people in danger because of the infectious disease's ability to spread to other areas via airborne transmissions. We should start a government program to enable more Americans to purchase or build a home different from past programs like HUD. Having a home enables citizens to quarantine themselves and avoid infecting others in a pandemic, when everyone needs to comply with health-care orders.

For example, the COVID-19 virus is coughed out through the respiratory system of homeless people, and the virus lands within the perimeters of a citizen's yard, doorway, window, or the ventilation systems of residences and buildings, which is a danger. When parents quarantine themselves and their children, teenagers, or grandparents at home, this doesn't guarantee improved safety. The testament is that our own humanitarian failure to provide housing can become our own demise during pandemics.

Worldwide, many governments have failed to provide decent housing for the needy during the pandemic. Government neglect for needy homeless communities has contributed a pathway for the SARS COVID-19 virus to increase its spread. Since state and federal governments have failed our communities, we need to end homeless encampments nationwide in America. Government employees haven't been acting daily to relocate homeless encampments to approved areas. Taxpaying citizens have seen no effort from government agencies or local governments to fix illegal occupations and trespassing on land owned by city governments.

In the museum, in a video, we can show how an ill homeless person coughing near homes where taxpayers have their children, exposing families to a dangerous virus. It's time to clean up America and end homeless encampments in every state, making these shanty communities illegal within city limits for health reasons. The pollution and fecal matter these homeless encampments bring to cities causes environmental problems as well as an increase in communicable diseases. Taxpayers shouldn't be exposed to communicable diseases by homeless people illegally trespassing and making encampments near their families' residences.

The American government has perpetuated the pandemic's spread further by refusing to obey professionals with the correct answers to resolving an international emergency. There are virulent viruses that are attacking the lungs and starving the respiratory systems of millions of people of oxygen. Pharmaceutical companies must produce an asthma inhaler that disperses aerosolized chemicals that will kill the virus on contact and not harm any other tissue. Aerosolized mists that can be sprayed into the upper respiratory track must be made as a first line of defense. Should a virus bypass a mask, the virus will come in contact with a solution that will kill it dead. We must stop a virus from entering the nasal passage quickly to ensure that health-care workers are well protected to do their job. Think of the aerosolized mist as a vaccine that is also a barrier.

The aerosolized pharmaceutical mist is one more layer of protection that prevents the virus from coming into contact with healthy tissues or mucus membranes. The aerosolized pharmaceutical for the nasal passage or mouth will have a new chemical chain structure that encapsulates a strong alcohol. The human body is unable to absorb the alcohol. The alcohol molecules remain in their spheres like an acidic mouthwash but without any potency to burn the mouth or respiratory tract. In its nanochemical structure, the encapsulated alcohol solution is capable of killing the virus on contact without causing harm to the human body. Nasal passage mucus can have living cells, and shielding these cells from a virus is one more preventive measure.

These types of newly engineered aerosolized mists can be used by physicians in hospitals who must constantly wear masks and PPE. The long time periods throughout the day when the PPE is worn can be uncomfortable, and where human error can occur, these pharmaceutical mists can help. Patients with a hundred multiplied viruses will be more likely to recover than, say, patients with millions of replicated viruses enveloping their respiratory system.

Reducing the number of viruses early in a pandemic starts with being prepared and having inventions that halt an infection's ability to replicate. Supporting inventors cannot be overlooked because the lives of those in our society will depend on the equipment they design for health care and survival. Research laboratory scientists and inventors need their own personal space undisturbed so that they can perfect their work.

Many medical scientists and inventors throughout history didn't possess professional credentials; they were just educated people passionate about finding truth. This museum will also acknowledge passionate professionals who attempted to aid the advancement of medicine.

Yes, for those who are reading this book, you have many brilliant contributions you can make to the advancement of humanity and the progression of medicine. Each generation has its own obstacles to overcome, and one of the best ways to help all citizens is by encouraging them to be creative in every capacity. We need to learn how to program the mRNA of ocean protists to attack and kill the COVID-19 coronavirus. The protists must be taught to eat specific strains of the coronavirus dangerous to human beings. Once the mRNA process is perfected, then we can replicate it so that new proteins battle infections.

Protists can be taught to battle many different diseases in biotechnology. An artist in medicine who spends time inventing can be a crucial component in finding a cure to many diseases. Medicine evolved from the arts, with many artists doing anatomical drawings while developing new inventions in their laboratories. Don't allow the current structures set in place for controlling medicine inhibit your ability to contribute in constructive ways.

Inventors must often work outside the boxes others want to encapsulate us in to make breakthroughs for humanity's future.

4

THE PYRAMID PROJECT AND COMBATING GLOBAL WARMING

The Pyramid Project is a fish farm artificial reef of stackable building blocks for various underwater aquatic species. One person with basic carpentry knowledge can assemble the structure by themselves with airlift bags, a boat, and scuba gear. The purpose of the wooden or steel mold is to generate interlocking concrete slabs. Skilled scuba divers, can erect the habitat for aquatic life needing a safe ecological environment. The public can participate in this project, adding their family's name, children's names, spouse's name, or any name they desire to their funded block. Some people may want to create an artistic design with stones and pebbles, by inserting eco-friendly materials into the moist concrete slabs. Stones and pebbles can be used to create names or art before the mixtures harden.

Love messages from newlyweds and couples can also be imbedded in the concrete forms, that are purchased for the project's development. Smaller vessels can stack the blocks, increasing efficiency, while minimizing fuel costs, and enabling easier material transport to the research site. The structure can be assembled on a flat surface in fifty to sixty feet of water—no decompression required—while other research sites can be deeper.

Erected underwater pyramids can evolve, in biodiversity interacting within its new aquatic biome. As ocean life adheres to the structure and organisms circulate throughout the pyramid, many fish species' can be reproduced while being monitored. Once one pyramid is assembled so coral adheres to the structure, additional pyramids can be placed nearby, expanding the research. Eventually, the pyramids can be interconnected with nets to encapsulate the site for further research, conservation, or fish farming.

The behavior of aquatic life, biochemical processes, and marine biological activity can all be monitored by science divers as adaptation occurs. Remote research locations can relay video signals from underwater cameras. Transferred data traveling up tethered mini buoys will be broadcasted through cellular phone communications or VHF and UHF radio frequencies. The mini buoys generate solar power for lighting artificial reefs acclimating to temperature variations.

As the world population grows, and feeding humanity is a concern for planetary engineers and environmental climatologists new changes must be made. Depleted fishery stocks in open oceans need replenishment. Research on Earth can be vital to expanding research on extraterrestrial bodies like the moons of Jupiter or the frozen volcanic tunnels of Mars. It's possible our future astronauts will be able to melt ice on Mars, thereby encapsulating fluid into volcanic spaces for producing aquatic biomes.

Martian aquatic biomes could be researched for subtle changes in microorganisms housed separately from one another in self-contained biomes. Inserting Earth's marine biology into planned Martian structures will create habitats like giant aquariums. Many moons and planets in our universe could be changed through terraforming processes. Being a master diver has inspired me to dive into the fluids of other worlds inside a special hard suit designed for extreme hyperbaric pressures in outer space. The extreme hyperbaric pressures from other worlds in our solar system provide an opportunity for new exploration. Space diving under frozen extraterrestrial surfaces and into diverse liquids from other worlds should yield the discovery of new extraterrestrial life.

I aspire that a single pyramid be successfully built. When one project is built, it could lead the way for more projects to be built along many coasts. Pyramids built underwater in Central America, South America, and even along the Old Jaffa port in Israel are a possibility. Marine life needs a peaceful sanctuary, and we can provide this.

The project would require a staging ground on a beach where concrete pieces could be stored and eventually moved out into the ocean, sea, river, or lake. The project requires wooden molds, concrete, rebar, and a vessel equipped with airlift bags to relocate pieces to the underwater erection site. Science divers would have to bring along their own diving supplies including breathable gases for the diving operations.

The versatility of these blocks would be helpful in protecting coastlines from oil spills and in the conservation process to restore wetlands along coastlines. It's possible these blocks could also be used in various construction barriers or the rebuilding of damaged areas needing improvement. My plan is to build one prototype pyramid with students and alumni from a commercial diving school.

The build site has not been chosen because I have not found one person willing to help me establish a staging ground on a beach to assemble the components for the project. One additional idea is for Israel to relocate materials from the Sinai Peninsula to erect small islands along the coast for new fisherman housing. Some of the newly erected islands could house families, resort lodges, and a marine sciences sanctuary for visiting international divers.

This idea is basically doing what is being done in Dubai, creating new islands and reclaiming waters for luxury habitation. The Israel project can include the improvement of aquatic biomes and cleaning up the seaports. The coast of Israel has great potential. It can become a series of self-made islands utilizing mountainous rocks from the Sinai Peninsula which are sent to the coast by railroad. These rock formations can be used to create artificial reefs for a variety of fish species.

In many cities, green-friendly restaurants can collect their food waste each day and send it to a processing facility where it can be converted into fish pellets. The fish meal is first boiled to a high temperature, then dried to make tiny nutrient balls that can be fed to wild fish on farms along coastlines, rivers, seas, or lakes. Fish large enough to be harvested for human consumption will be processed, and fish scraps will be sent to farms or greenhouses as fertilizer. These fish farm aquatic biomes can create a cycle to help sustain human life in coastal communities.

Keeping fishing harvests free from pollutants is invaluable to all people. Many island nations can collect uneaten nutrients for processing into commercial fish meal pellets. Around the world, human beings need to learn how to stop being so wasteful and to recycle food scraps for commercial fish meal pellets. The fish pellets can help increase the nutrients needed to establish an open-ocean fish farm that yields greater harvests. The development of artificial reefs benefits many marine areas that have been depleted of new schools of fish. Artificial reefs also create new diving opportunities for international travelers. Many oceanic areas have reduced fish populations, and we need to redevelop the stock.

Ocean life benefits from new habitat creation through artificial reef development. Artificial reefs provide attachment habitat for new life to generate, absorbing carbon dioxide to reduce global warming. Artificial reefs provide research opportunities for scientific divers monitoring environmental changes due to global warming. New artificial reefs will provide new opportunities for divers to begin cleaning up waste when it's discovered. All divers must do our part to take care of our planet.

How much money do governments around the world spend on cleanups after natural disasters? Many islands have been destroyed by: hurricanes, tsunamis, typhoons, cyclones, floods, and tropical windstorms. So many coastlines and inland territories have been destroyed by severe weather washing hazardous waste into our oceans' ecosystems as pollution.

Marine organisms have been combating humanity physically by trying to adapt to all the harmful chemicals swept into their environment. We need to make a change for our world's environment. What we need are people with abundant resources allowing us to use their marine vessels and equipment to do the ocean cleanups. There are thousands of yachts and research vessels anchored at marinas or ports which could invite scuba divers aboard to participate in ocean cleanups. Our waterways need all the trash to be collected and to have it sent to recycling facilities.

The Sea of Galilee in Israel can host new artificial reefs and be an opportunity for environmental cleanups leading to a (100) one hundred percent improvement in water quality. The entire perimeter of the Sea of Galilee can be a project area to cleanse the lake of all impurities through marine divers doing biological enhancements. The goal will be to transform the Sea of Galilee back to the way it was thousands of years ago and to beautify all surrounding lands where Jesus Christ ministered. These environmental projects to get rid of the pollution in the Sea of

Galilee will start movements in other regions to detoxify water and eliminate waste problems.

The Black Sea, the Caspian Sea, and the Dead Sea have an opportunity to conduct terraforming research. We need more artificial reef processes through dedicated experts making development improvements. The world's population is steadily increasing to nine billion people all needing to be fed. People must become prepared for the future. Israel can pioneer processes helping to ease mankind's burden before famine or pestilence emerges. Only through a combined effort can real change occur. It's much easier to complete large projects with teamwork rather than alone. If a scuba diver wants to construct an artificial reef alone with their own boat and diving equipment, it can be done with our fish farm building blocks.

A nation that works, is a nation that eats, for when misfortune arises, the preparations they've undertaken are a testament to their wisdom. Whether it's the ancient storage houses of Egypt or modern-day silos in farming communities, our nations all need to eat for survival. Wise nations plan for the future by taking action before huge problems start surfacing, creating national emergencies. Nations that are not wise ignore the truth, and when tragedy strikes, these are the first people to develop strife in their territories, leading to violence.

We can provide a schematic for how to erect food pyramids with all the necessary information to bring awareness to these current and near-future catastrophic problems. These food pyramid schematics can be placed in dive shops and schools in a variety of different languages. We're an environmental global coalition united for clean water, tree planting, Earth's preservation, and artificial reef development, working to protect all water ecosystems for humanity. Clean water is needed for all humans, animals, marine creatures, and unique organisms needing purity in their environment to maintain life.

All countries deal with human waste daily, so we must develop better sanitation processes to transform waste into usable organic fertilizer compounds for farming and trees. To keep water cleaner, we must address all problems areas that create pollution in the water by implementing daily work objectives to reduce the number of pollutants entering the waterways. Students must begin projects in classrooms, creating new designs to fix environmental waste problems. Students must create work lists to reduce water pollution.

Students must go out on school-organized field trips that host a competition to see who can collect the most pollution, waste, and litter by weight. Unknown to the students who don't take the competition seriously, an important public figure may present a prize to the student who did the most work. If the person giving the prize is a billionaire saying they did a great job, that's inspirational for all the other students.

Holy sites globally should be some of the cleanest lands, for when these sites were created they were pristine, and we are the caretakers. Israel needs to be cleaned up with a daily program of revitalization to its environment. Out with the litter and pollution so there can be new landscaping with irrigated flower beds, shrubs, and newly planted trees throughout Israel. We can hopefully provide monasteries, seminaries, convents, and churches in Israel with the resources they need to maintain the work of keeping the nation beautiful.

One proposed scientific experiment is to have a group of volunteers stretch a thick, clear plastic tarp over a landfill that processes waste on its grounds. Over a forty-eight-hour period, the tarp would collect gases escaping from the landfill into a deflated weather balloon attached to the center of the massive tarp. After the forty-eight-hour period, the inflated balloon would have all the gases collected for funneling. The funneled gases from the balloon would be fed into a two-process analysis using column chromatography and a gas chromatograph. Theoretically, gases from landfills in all countries around the world, when condensed in the upper atmosphere, are contributing harmful chemicals to all ecosystems.

Human beings, animals, and all living species are being affected by the chemicals introduced into their environment. How do people develop strange diseases or unexplained illnesses? At the bottom of landfills, we have black, oozing liquids full of chemicals that are also rising into the air. Science will reveal the toxicity concentrations for each landfill that are mixing with our upper atmosphere. Three hundred sixty-five days a year, toxic gases are rising from landfills worldwide into our atmosphere.

The data from this environmental research on landfills will provide projections for years to come on many factors contributing to global warming. The need to create domes over landfills and to collect all the gases for proper processing will be another potential contribution society can make to improving the health of our planet. Landfills, I suggest, need to be under bladder biospheres where the temperatures cooling the waste below and vapors seeping upward are collected and processed differently from natural-occurring gases.

Many man-made products contain chemicals engineered to behave a particular way, and they shouldn't be mixed within the air that populations around the world breathe daily. It's not just the smokestacks from industrial corporations worldwide creating pollution. An effort should be made to improve all areas of sanitation in every form on a global scale. Recycling facilities utilizing prisoner labor (365) three hundred sixty-five days a year is a start to reduce waste and help improve the environment for the growing worldwide population, now near nine billion.

Reducing poisonous gases in the air is mandatory for scientists and health care professionals. Many diseases are linked to exposure to liquids and gases, which create physiological problems not just in humans but also in aquatic life. Gas-liquid-partition chromatography (GLPC) or vapor phase chromatography instruments must accurately measure the entire volume of gases in the proposed experiment.

The variety of gases collected in different countries can also be valuable data, showing the different concentrations and types of chemicals being released in the upper atmosphere that mix with chemicals from other countries in the clouds, creating smog. Ozone depletion must still be monitored, as well as acid rain, in order for nations to get a handle on the dangers to our environment. It's not just the microplastics entering our oceans harming fish and wildlife. It's our mismanagement of our energy systems, for solar power needs to be combined with wind power to make windmills that are also solar-power-collecting blades, cutting expenses.

On an ecological scale, the chemicals in our atmosphere condense into liquid form, mixing with ocean water, drifting onto coral reef biomes, and coming in contact with various aquatic species. The life we consume from aquatic biomes needs not to be exposed to these chemicals from our upper atmosphere. Humans are not the only consumers in the food chain; there are many other species ingesting chemicals they did not ingest before humanity industrialized. As a world population, we must all be responsible for ensuring that we have the means to safely feed humanity in the future.

It's our responsibility not to expose humanity to deadly toxins airborne in the environment or in their food. People can easily get ill from food-borne pathogens and even die from E. coli, salmonella, botulism, amoebas, enterotoxins, tetrodotoxin, viruses, parasites, and bacteria. Protecting our Earth and caring for its health are our responsibilities toward our Eden, our home. The diseases humanity is experiencing have an impact on the environment. Human diseases have been exposed to many creatures worldwide through pollution or close contact. Many of the creatures hosting diseases can become incubators, mutating diseases that, in turn, return to humanity via our ecosystems.

The respiratory illnesses creatures develop could have originated from the chemicals humans exposed them to through the environment. The viruses, bacteria, and parasites in these creatures' natural environment are now deadly to humanity. Humanity has said to the ecosystems of the world that they should adapt to the pollution now; the creatures of the world in these ecosystems are sending a message back to humanity to adapt. As the entire world is affected by the COVID-19 pandemic, we must ask whether we brought this on ourselves through our own arrogance.

The processes we perfect here on Earth will also help us in the colonization of Mars, Jupiter's moons, and beyond. The last thing anyone wants is for humans, aquatic life, and organisms to be affected by a new disease that adheres to the chemicals in our bodies, particularly those created by mankind. There are so many diseases and cancers without cures; therefore, as a human race, we need to work together to clean up our environment.

Our nations can build massive cities with transportation systems. The focus now must be on our environment, with streamlined railways funneling waste to massive self-contained recycling facilities employing prisoner labor. Every piece of trash and all material entering the recycling facility must be broken down, reducing landfill accumulation. Reducing the stress on our environment by recycling everything helps the longevity of our natural resources and reduces the depletion of less-renewable resources.

We also need to funnel human waste out of cities more efficiently via sewer tunnels beneath cities. Millions of toilets flushing need their contents funneled into sewer reservoir settling pools underground, which then funnel septic water down over waterwheels, generating electricity. We need to make the sewer system work for humanity, providing energy to our cities. Also, we need to get diseased fecal matter out of cities in a more efficient way to reduce any spread of infection. The faster we get feces out of cities, the faster we prevent rodents as well as other life from being exposed to infectious content.

The bubonic plague is just one example we don't want to repeat through poorly established city infrastructures. Also, human fecal matter should be processed through prison labor in a series of scientifically proven processes for composting and waste management. Reducing the amount of human waste entering our waterways after storms, hurricanes, and natural disasters means having a well-planned method of diverting waste out of cities to prisoner compounds.

Prisoners require punishment for their crimes, so it's a necessity for them to labor for the human race, by preserving the environment which sustains all forms of life. Prisoners who harmed human life must labor in environmental sanitation as a form of punishment. Young people on the streets of America need to hear that, in prison, they'll have to process recycling waste and human feces. Hopefully, this information about prison sanitation will convince young people not to become incarcerated and not to obey drug dealers.

I would like to be elected president of the United States because I believe that, during my term, I'll make positive changes for the entire world. I know that as president, I wouldn't be able to clean up all the problems in Washington, DC. At least during my Presidential term I will try to clean up some of the Washington DC corruption. Those who follow in my footsteps will have less of a mess and can continue to do good works towards a peaceful global progression.

Abused prescription drugs and illegal narcotics remain a global pandemic. In today's society, too many adults are drug users and alcoholics creating problems within our nation. This nationwide corruption needs to stop, with state and federal agencies making hard decisions for all the children of this world. During my term as president, I will avoid all forms of war because nations around the world already know the United States is a superpower with the ability to initiate war. It's time for us to look beyond war and make positive, constructive changes in the way we work with other nations around the world.

The congressional design of the past cannot be the congressional design of the future. As the world changes, so the US Congress needs to change in its methods to serve the people better. As a scholar, it's my duty to help perpetuate the progress of humanity and to work with others within my administration for this purpose. Children worldwide are the future of all nations, and they need the most honest leadership guiding them. It's the children of this world who will follow in our footsteps to continue our humanity throughout the universe. We need to support our children's technological achievements in all forms of industry and science for our space program's future. One of my goals during my term in office is to work for education for the advancement of technology.

Internationally, our students will continue to advance scientifically through engineering and mathematical programs, moving forward our progress toward the colonization of Mars. Today's youth in the United States need to learn the importance of protecting the environment and cleaning up their communities. When Americans arrive on Mars, it will be our duty to ensure we don't pollute another planet like we've polluted the earth. We must become experts in sanitation, recycling, and all environmental issues before reaching other worlds.

During my term in office, I'd like to support programs in education that help teach young people the importance of caring about their environment. Beautifying America (BA) will be a program for all students nationwide to learn about becoming green friendly outside the classroom. Beautifying America will be a way for students to earn tuition for college while working toward a college degree. Students working in their communities in the BA program will begin to earn credits toward a bachelor of arts degree.

Members of my administration will communicate with teachers in all school districts, working with them on this national program to become green friendly. We will get our youngsters out visiting national parks with teachers and participating in field trips to enhance their developmental health. The entire nation will benefit from the Beautifying America program. Students need to learn why it's important to care about their environment by participation in scientific field studies. We will enable students to interact with biology in their communities, increasing their scientific knowledge about the nation where they live.

The study emphasis for students will be environmental science, zoology protection, conservationist studies, hazardous materials cleanup, fire prevention, biology, botany, forestry services, and pathogen prevention. During field trips, students will work together to clean up their environment. Students will also tend to wildlife harmed by human waste through hands-on programs. Students will be taught marine biology and ecology as part of their mandatory fieldwork during their science class excursions. In the classroom, students will be tested and given the opportunity to voice their concerns.

During my term in office, I will focus on programs within America. I will also tend to the needs of Americans first before dealing with any other countries' requests for aid. Stopping the homeless encampments nationwide is one way to reduce the spread of infectious diseases. During my term in office, members of my team will provide better housing, health care, and meals to all Americans in need. Together, we will clean up all homeless encampments as we sanitize and clean up our nation to beautify America.

I believe I would be capable of doing the work of the presidency of the United States because I enjoy office work while serving our Lord Jesus Christ. I'd try to make the best decisions in the most positive light as to how Americans need to be seen through fruitful works of kindness. We need to be kind to our own citizens and to improve the way we treat each other to stop violence in our nation.

Stopping problems in America starts with electing me to the office of the presidency of the United States for two terms. Why do I desire two terms in office? The first term in our nation's capital is realistically a term of learning the ropes to deal with hardliners all wanting their political way. The first term in the Oval Office is hard for any newcomer because of all the duties that this office requires. A mega-skyscraper construction project can easily take many years. Politics isn't much different from construction projects because they both require social bridges, alliances, and new foundations to be developed to get work accomplished. When we allow people into America, we expect them to work hard because this is what America's working class does every day.

We're going to stop organized cartels from other countries from sending gang members to spy with soulless eyes, seeking out any angles to undermine our nation's prosperity. As president of the United States, I will be the least merciful of any American president when it comes to drugs harming our nation's civilians. Locking up drug-dealing gang members, cartels, and families participating in illegal drug network activities will be a hard-line objective during my presidency. If I must use the military to lock up all the criminal drug dealers across this nation, I will do it. As president, I must protect our nation's children from those drug dealers who want to poison lives and fill our morgues. We will go to war with drug dealers like never before.

I will send the strong men of our military to hunt down, subdue, and imprison anyone involved in illegal drug distribution. The drug dealers will resort to violence and have been resorting to violence for a very long time. The difference between the drug dealers and my administration is that the task forces I establish will be the most elite, able to crush their entire empire in one swift campaign. If we need to imprison a million drug dealers, then we will imprison a million drug dealers. I will build a massive prison in the desert mountains, we will corral the criminals into one big city, and then I will work them to repay for all of it, using military punishment.

The technologies we have today can find every single drug dealer on the planet in an instant, so we can stop their tyranny once and for all. Like a powerlifting Olympian raising the heaviest weight, we will snatch up the drug dealers and cast them into prisons for the remainder of their lives. Drug dealers are terrorists, and their murder of innocent lives and fighting law enforcement are coming to an end.

Our genius American scientists will use a new herbicide to destroy fields of illegal drug crops. The war against cartels will be an all-out war to crush their empires. Cartels will shoot, bomb, and attack innocent lives, which is terrorism, so our military will wipe them out. We will send our new hand-selected drug-enforcement marines to strip away drug dealer wealth and put them out of business. Hundreds of thousands of lives have been lost because of illegal drugs, so our marines need to wipe out hundreds of thousands of drug dealers to balance the scales.

We will give our marines the most advanced, state-of-the-art weapon systems to catch the criminals like ghosts, snatching up cartel bosses who hide, thinking they're safe. The humanity of the world is not going to be dictated by a bunch of losers with second-rate educations who sell drugs to ruin lives. All the money of the drug cartels is going to vanish so that those who do survive will live in poverty like the families they've oppressed for decades. Right now is the time to record every single drug dealer on the earth and make plans for the future to end their tyranny, stripping away all their resources.

After the criminals are locked up during a period of difficult times, then there will be a period of progress when those who are smart can show us what they have for space development. In the near future, corporations around the world will have to work together, and many advancements will be made toward building an outpost on Mars. All it takes is one billionaire with a focus-driven plan for successfully placing our first candidates on their Martian outpost. There is always a more economical way of doing things because, as innovators, we create what others say we cannot create.

So now I'm going to share two projects I believe are necessary not only for global warming but also to test an approach to fixing a greater problem. I believe what we do here on Earth can influence what we do on other worlds, especially in science and technology. Sometimes we will have to change designs along the way, but as long as we test our hypotheses, then we aren't aiming for failure. Data is data, and rock-hard data gives us the foundation we need to formulate a feasible plan for inhabiting other worlds.

Sometimes it's the innovative people who were never invited into the inner circle, board room, agency, or project who can contribute the most. These innovators are valuable because they bring new, fresh, and different ideas that haven't been presented to teams for analysis. One way of thinking isn't always the best; as innovators, it's your job to have multiple ways of thinking to keep your mind open to new information. Sometimes there are people working on teams who cannot accept that they didn't create a brilliant idea, and that can create conflict.

Beware of those who are overly self-centered, for it's very important to hear what others have to say. Yes, we have rovers on Mars collecting data. Did anyone suggest designing them with easily swappable batteries to enable longer usage? In the aforementioned example, we will spend millions of dollars sending a robot millions of miles away from Earth to another world, but let's get the most out of the investment. An idle robot isn't collecting data, so when we send the next robot, teach it to swap out batteries. Battery improvements are just one way to keep mission teams gathering data for the next mission, in addition to repurposing resources to perform other tasks.

Here is a plausible science project that could be built within some locations here on the Earth. Millions of years ago during the Cretaceous period, North America had an inland sea teeming with life. Today, these American inland seas are dry areas of desert along our highways with vast regions being built into emerging communities. I'd like to create a interconnected series of mini inland seas, three of them utilizing siphoned seawater from the Pacific Ocean or the Gulf of Mexico into each irregularly shaped canyon repurposed for planetary engineering. The man-made inland sea would be like restoring the ancient sea to its original place to promote new life. Scientists talk about rising ocean levels due to global warming, but what about diverting some ocean seawater into new man-made inland sea?

America is an innovative nation. Americans have created golf courses, oasis resorts, winding marina retirement communities, metropolis cities, and space programs so re-engineering earth's landscape for scientific terraforming is doable. What we learn here on earth we transfer over to other planets mastering new skills in planetary engineering. Changing the landscape and ecological functions of extraterrestrial planets requires precise scientific data. Well, let's get some practice terraforming our own world first to help us learn ways to better terraform other worlds. Meteorites that struck the earth millions of years ago, I believe, had a contribution to destroying America's inland seas. The impact of a meteorite fracturing the earth could have given just enough space for inland seawater to escape in valleys between mountainous ranges while also vaporizing seawater during the explosive collision. In the meteorite impact some seawater vaporized instantly, while other areas with seawater were shielded by sandblasted mountain ranges becoming dust pushing trillions of gallons of water out like a tsunami flood of silted mud. The muddy seawater from the meteorite impact entered the Pacific ocean and the Gulf of Mexico settling on the continental shelf. Ocean currents have moved sediment from the meteorite impact further down into the depths of the benthic zone. Transforming the South Western United States into a new inland sea with marina communities inside a science zone operated by professional billionaire scientists funding the project is the mega project.

What I'm proposing is a process in which we use differential pressure to move ocean seawater upward, spilling out into a series of irregular-shaped bowls of canyon territory, each having its own depth. Like a siphon, we drill deep into the earth, creating a tunnel that extends out into the depths of the ocean beneath the bedrock of the coastal shelf. The tunnel isn't breached until the last moment when we want the seawater to flow into it, filling the space we created. There is no hydraulic fracturing involved this boring tunneling project which is a very precise planetary engineering plan with scientific methods followed to achieve optimum results.

The seawater will travel down then upward into a narrowing passage to where we want the water to flow. The massive pressure from the weight of seawater above the chamber tunnel aids the upward flow to our project filling our three interconnected mini seas. The weight of the seawater above the capillary columns of tunnels will create a bladder-like effect, moving seawater into our inland seas in a circulatory process. The daily tides that take place on our oceans will keep seawater moving so it doesn't become stagnant in each of the inland seas. Seawater in the first inland sea will be at a higher elevation and will run down to the other two shallower inland seas, each with a different depth. In the three tier level system, seawater will be siphoned up to the mountainous heights and then be funneled back down again to the ocean through a descending rotational flow of cascading rivers, splintering streams, and past our mini hydroelectric generating stations. Streams should be a variety of widths, shapes, and sizes, changing seawater flow in dynamic movement for desired currents aiding aquatic species adaptation to the project. We'll develop many mini hydroelectric generating stations for an unlimited supply of electricity in this project.

Think of a series of rocky waterfalls moving seawater containing wild fish from breeding farms down oxygenated waters through a step-down circuit. The seawater will return to the ocean it came from with trillions of wild-raised fish that were kept protected in no-fishing seas. Of course, as the new inland seas become inundated with life, more carbon dioxide will be absorbed from the upper atmosphere. I've looked at two possible locations for this project, having traveled across the United States numerous times. Southern Texas near the border with Mexico is one possible location for this inland sea project. The second possible location would be Southern California and Arizona between I-10 and I-40. The region near the Mojave National Preserve has mountains with different elevations that could encapsulate vast amounts of seawater. The mapping of all the regions where we can actually acquire desert land for this science project is a surveyor's zenith to paramount. Transit level technologies which analyze topographies using drones and imagery calculating different forms of geological structures will make this work tedious for it's massiveness. What are the benefits for the investment of time and resources in this science project? What will we reap from the pyramid project: sustained electricity, aquaculture food, commercial fisheries, agricultural farms collecting freshwater, enhancements in ecology-conservation, new marina retirement communities, hundreds of thousands of jobs, and enormous data for planetary engineering. There will be financial profits much greater than the initial investment in this planetary research for inhabiting other worlds. I believe this science project with it's team of first investors could become the first trillionaires in the world by acquiring new patentable scientific technologies. If you want to be a trillionaire you have to think like a trillionaire.

The man-made inland seas must have four different depths in each section of the three adjoining and adjacent quadrants. We must try to mimic what inland seafloor topographies already look like so that our project will be adaptable for fish and marine life introduced into the project. Coral reefs have been destroyed in other parts of the world; we can try to restore them in these three inland seas. Creating a preservation habitat untouched by commercial fishing and human pollution is beneficial for our oceans. The three sea lakes are to be very large and won't become endorheic basins closed off without water flowing out.

The three inland sea lakes are to not be anoxic, thereby moving salt water from the ocean back to the ocean. Ocean seawater will recirculate moving from high altitudes back to sea level through fluid dynamic designs for generating hydroelectric power in interconnect mini plants. These utility power designs will create circuit transferring energy through seawater movement flowing down smooth channels turning a series of power turbine sea water wheels. States like Texas, New Mexico, Arizona, Nevada, and California can benefit from the hydroelectric power stations operating in the evening time. Seawater is not uniformly saline and isn't entirely sodium-chloride based. This information is important because, on other planets, all these variables will account for significant changes in our terraforming methods. We do not want to have to regulate the amount of salinity in our project as we strive to keep it all natural. The process of extracting salinity from seawater will be discussed later, with a potential method to give all nations unlimited fresh water from our oceans.

Speleology can be part of each of the inland lakes, making cave systems for different marine life to have sanctuary and for scientific divers to monitor the project's advancement. In the inland sea project, we should have river siphons when seawater passes under submerged objects and around boulders. Designing rivers with turning channels is part of the engineering process to make a very well-thought-out sea that becomes (100) one hundred percent real, functioning on its own. When we terraform other worlds, we must develop processes that are self-governing, regulating themselves in our absence. Marine creatures reproduce more regularly when they feel safe in the natural environment in which they live.

Self-siphons would be beneficial to use in some spillover dam sections that can generate electricity. Some waterway channels of the man-made inland sea can have seawater waterwheels extending over river currents to rotate frequently generating electricity through the use of mini turbines. Like suspension bridges smaller cables attached to tethered steel poles will span from either side of the waterway holding up the cylindrical shaft crossing the river, stream, or aqueduct. Beside the seawater waterwheel shaft mini turbines shall align the embankment of rivers being interconnected in a series circuit to transfer power to transformers for electrical distribution.

The horizontal waterwheel shafts extending over the river would be adjustable, rising up and down being lowered in slotted notches depending on the depth of the water current flowing. Strong shackles used in rigging will attach midway and periodically along the shaft spanning the distance across the river for cables attaching to welded eye holes. The generating of electricity over rivers and aqueducts covered with rocks moving water rapidly also has room for improvement in freshwater operations. A secondary project is taking water from larger lakes to make a series of smaller interconnected lakes. Both the freshwater inland lakes and the saltwater inland seas have merit for gathering data that can be used to terraform other planets. Freshwater inland lakes and the saltwater inland seas need to manage the colder temperatures of other worlds for flow dynamics.

Our solar system has planets with volcanic activity that can heat regions differently. Some planets have areas with frigid water beneath ice, frozen land masses, and extraterrestrial frozen chemicals we must account for. If we conduct our inland sea projects in colder Arctic regions as well as in hot, arid regions, this information can be invaluable, giving us a barometer of data for terraforming planets. We must be very accurate when dealing with oceanic microbial components in each depth of our inland sea project.

All depths of our inland sea must be dived to monitor the thermocline for any microscopic variations while divers take samples for lab analysis. A census of marine life in our project must be taken regularly to record any and all changes in the habitat we've created. Microbes on Earth may behave differently when transferred to outer space and onto other planets for establishment in new environments. I'd like to have two independent teams working on this project if and when funded with all approvals to go ahead with the work.

Each team of experts shall review the other's work and shall consist of ten professionals. The ten professionals having accredited degrees, and passionate for science shall provide their research. Ten people can nicely fit around a conference table, providing their valuable insight. The two teams will operate from two research buildings opposite each other across a canyon. Some professionals will have overlapping interdisciplinary degrees, which will be invaluable in making this experiment a success. All the scientists will use scientific methods with the most stringent rules.

Each team should consist of: a planetary engineer, a marine microbiologist, a geologist, an ecologist, a toxicologist, a botanist, a zoologist, an environmentalist, a conservationist, and a geneticist. The two teams, working separately, will prepare documents based on their research for review by others in their own network. It's possible that the two teams can come from two separate universities. Each team will be able to hire scientific divers, commercial divers, aquanauts, and additional crew members as needed to aid them in completing their research.

These scientific teams are necessary for space exploration and the development of other worlds for the continuation of the human race beyond Earth. In addition to the ten professionals on each team, there will be one medical doctor of space medicine assigned to each group. Periodically, the medical doctor will monitor and evaluate the research as well as the teams to ensure they're in good health. Diving teams will be submerged frequently in warm or frigid waters, conducting scientific experiments and performing their research. The ecological analysis of organisms and how they thrive in the natural environment that we created for them will be monitored.

There will be the opportunity to build an assortment of underwater scientific research habitats at depth before the canyons and mountainous regions are flooded with seawater. The development of these strong habitats before flooding a region will keep the construction of an underwater base more economically feasible for our aquanaut science divers. When two entrepreneurial billionaires each funded one team, then this project can be built once the property and permits are allocated with government approval. America likes big grandiose projects like building pipelines and roads for thousands of miles, so why not a subterranean tunnel deep into the ocean creating a new power source?

Often it's very difficult to get any project approved in the United States due to its many regulations, laws, and jurisdictions all needing to give permission. Many American companies have sought to develop their ideas overseas when they've encountered too many denials. It's possible China, Russia, Africa, Europe, or quite possibly a Middle Eastern country with mountainous deserts could be interested in making this project if America isn't interested.

Our outer space terraforming projects of other worlds must have homeostatic environments for all forms of life to flourish. On extraterrestrial worlds, we may need to establish new aqueduct technologies, cisterns, and terraforming processes to aid a colonies' survival through backup systems should a well-thought-out design fail. All space missions should always have two additional systems for survival using analog processes should the digital systems fail.

Presently, there are freshwater shortages around the world and regions with droughts. How do we get the fresh water necessary to meet the demand for growing crops, and also providing all living species' they're needed daily consumption? We need fresh water for our daily consumption. The answer is let's look to our oceans and redevelop a set of new processes for harvesting fresh water from salt water.

Think of a process in which a tea bag is a special ion compound that attaches itself to salt molecules, solidifying them. We insert the tea bag into the bowl of salt water and raise it up and out, discovering the entire tea bag has solidified into a salt block. Repeat the process, inserting the tea bag again for good measure, and then test bowl's contents for any salinity. If there's no salinity in the bowl of water, then great—we have consumable fresh water.

Now what do we do with all the salt we have extracted? Can we make the salt into a fireable glaze by mixing it with other compounds? Can we make the salt into a mortar-like concrete that can be fired into a brick for construction? Ionic liquids and molten salts are solid at standard temperature and pressure. We will need nuclear power on other extraterrestrial worlds. Hopefully, we can transform our new extraterrestrial world into a planet having all the colors of the light spectrum needed to establish a homeostatic balance. When we encounter extraterrestrial life in other worlds we may have to build around the habitat of these creatures to not disturb their homeostasis environment. How we learn to terraform the earth will be crucial in how we will need to terraform extraterrestrial worlds that host alien life possibly susceptible to our biological needs. Designing around alien life isn't much different than designing around the life organisms we already have in this world with billions of people. Zoologically alien organisms will have a different genetic makeup and new ecological processes which we will need to map around for our homeostasis living amongst them. Humanity contaminating alien species in their natural environment must be prevented. Also preventing humans and the earth species from being contaminated must be a priority when inhabiting extraterrestrial worlds. Extraterrestrial alien diseases have had a very long time adapting to the environments containing them and for this reason their potency is as strong as the worlds in which they live. The earth's gravity, temperatures, pressures, lunar cycles, chemistry, and planetary cycles can all play an important part in how diseases adapt or thrive. In extraterrestrial worlds we have to account for diseases having long periods of time to mutate or change based on all the factors influencing their development.

AUTOMOBILE COMPANIES WERE DEVELOPING ceramic engines, but what about ceramic water pumps? Salt water corrodes metal quickly, but a ceramic water pump could move fluids with more longevity, reducing the expense of replacing damaged parts. We often see natural gas companies and oil companies burning excess amounts of flammable gas from pipes extending out of the ground. What a waste of energy. These gases could be used to heat seawater and funnel steam into condensation pools, providing fresh water to many regions.

The oil field fumes wasted for decades added to global warming could have been heatingsteel instead thereby transforming constant supplies of salt water into steampower. Pressurized steam that's funneled as propulsion toward a turbine will generate electricity from all the burning gases that oil fields and gas fields vent. These plant utility processes are patentable once all the intricacies are precisely engineered for the most advantageous outflow. Geothermal energy has many advantages because the power is there, just waiting to be tapped into, even around volcanic vents. Using the earth's internal heat to evaporate seawater and provide steam to help generate electricity is how we need to think. Hydroelectric power designs of the past are outdated across America, and new designs need the implementation process initiated.

If we can perfect plant utility processes here on Earth, we can transfer these designs to other worlds with the necessary modifications being made once were established there. Remember many corporations including the government will often say "NO" to many things you desire, but don't let them deter you from progress. It's easier to make progress when someone gives you a "YES," but when you're dealing with people who give you a "NO," more often than usual, it's an indicator of another problem.

Politics in America can make or break anyone. So many billionaires stay away from politics because to them its a waste of time and I was one of them. My bank account didn't reflect billionaire status because I was stolen from before I was able to accomplish submissions of my patentable technologies which would have made me wealthy. This is why I now I find myself now immersed in politics but still clinging to business living in a median rather than my initial route which was altered by greedy family members thieving the entire volume of works from my hands. If someone doesn't fund your research or projects, then go to those who will, because this is the nature of business. If someone defunds you, then go to those who will fund you. If someone doesn't like you, well, then seek out those who do. If your family members are the embodiment of corruption then separate yourself as far as possible from their predatory behavior wanting what you have.

If America doesn't give you the success you were looking for, then do what many American companies have done and go to other nations. Many other nations will make the products or designs that you've created when you're able to establish effective communication with them. In the private sector the internet connectivity and it's security hinderances have made it difficult to communicate with businesses overseas. America's government has gotten way too controlling over Americans trying to communicate with foreign companies via the internet for business. What is happening is that large scale corporations are the ones getting all the communication access to foreign companies while everyday citizens with ideas are being blocked out from smooth connections.

If foreign governments want to do business with American citizens then they must send their representatives to America to meet with innovative citizens trying to get products made. The American government has gotten way too controlling to want a middle man position in internet communications and has teamed up with big companies to block out the little people in business. Many honest Americans feel very cheated financially by all the angles of corruption coursing through the veins of criminal Americans who are employed in government. Prove these allegations is the angle the criminals who are united thrive in, hiding behind their status, uniforms, and credentials caring nothing about others. The criminals are living a high life having with riches they've illegally acquired, while no one investigates them because the investigators are in on the crime obtaining wealth not upholding the law. So my advice to you is get away from the entire network of like minded individuals who are accustomed to living in a take everything from others lifestyle.

Jealousy is one of the biggest problems holding many inventors throughout history back from making breakthroughs or success in their life goals. True innovative people know others like themselves because often were the ones putting in (80) eighty hour work weeks not enjoying any fun! Innovative people work on their passion while others party away their lives. Many of the partiers are the ones who stole valuables from the innovative victims till they could get themselves in a position to be recognized socially. Taking away recognition from innovative inventors who work hard daily and who are not having any fun is enjoyable to thieves making plagiarism a hidden profession. The thieves enrich themselves from tons of information they've stolen in home invasions and live in a realm where there aren't any repercussions for destroying victims lives.

Don't let others devalue your hard work, talent, skills, and ingenuity by being condescending. Some people just don't want to see you financially successful. Some people just cannot accept that your the one with more talent being more focused on your goals than they ever were, so they want to sabotage your empire in they're rebellion. Often, these types of people will do everything they can to prevent you from obtaining an improved, financially comfortable existence. Many people cannot accept that they have mental illness and that the games they're playing with people's lives is a part of their delusional mental illness. We do live in a world of racism, discrimination, bigotry, hatred, and jealousy festering in the minds of people who want to oppress others. Rise above the minds of those who aren't aiding your productivity. People will say that such science projects are a waste of money and time, but I say exactly the opposite. The scientific research conducted in this pyramid project will help save humanity giving us invaluable biological data to live off this world. Fishery scientists will breed trillions of fish in inland fish farming practices returning them to the oceans and port facilities aiding to feed humanity in this project. Also, scientists working with pharmaceutical companies can raise certain marine species in a controlled aquaculture environment that will produce healthy organic food, cancer medicine and other drugs to help save human lives. The teams of professional scientists and diving crews will be biosphere researchers who gather important data on global warming while reducing its effects.

Deserts that were once lacking fresh water will receive it from the evaporation that takes place over the mini inland seas. To combat global warming we need massive grandiose projects because the effects of global warming are massive. Hurricanes and storm flooding can wipe out entire cities in many countries around the world. The earth has a huge upper atmosphere and to balance the equation we need massive projects on the ground to establish a equilibrium of gaseous exchanges. Yes my project is a massive undertaking but thats exactly what we need to prevent rising ocean levels and to redistribute some of that seawater elsewhere. The rising seawater elevations will effect the Thermohaline circulation within our oceans along with it's salinity so why shouldn't we engineer inland seas to move seawater around? Reducing the effects of hurricanes my pyramid project can accomplish because we will move seawater to deserts away from vulnerable coasts. The devastation caused by huge hurricanes has already crippled economies in many countries and the aftermath is terrible. We can halt the effects of global warming by implementing the designs professional scientists are willing to research with government endorsements for the pyramid project's development. Together let's begin transforming some unused desert regions into oasis's full of life having new marinas, seas, and living zones for aquatic species to thrive.

Desert plant species will be able to utilize atmospheric water vapor to sustain growth, collecting condensation in the evening. The wildlife in the desert relies heavily on plants for their intake of moisture. Animals benefit from water vapor that's available on plants for consumption. Salt water converted to fresh water provides trees, grasses, cacti plants, and flowering species life! Wildflowers draw honeybees, butterflies, and moths to deserts needing the crucial cross-pollination. Many billions of dollars are spent annually by governments recovering from global warming catastrophes. Billions of dollars are spent on weapons of mass destruction, which I consider wasted money.

The research from these inland sea projects will help generate new terraforming techniques. New designs will emerge for creating subterranean biosphere ecosystems for the Mars Pyramid Project. Imagine sophisticated tunneling equipment creating a very large step pyramid with vertical, horizontal, and sloping facet tiers extending outward beneath the Martian surface. Colonies are established on each level for each tier of the underground pyramid. Surrounding the pyramid on all sides are waterfalls, ski slopes, lakes, ponds, tributaries, botanical gardens, and agriculture biosphere tunnels. Think of the underground pyramid like a series of massive skyscrapers stacked side by side, making a step pyramid shape. Chiseling out skyscrapers beneath the surface of Mars in precision can be the modern day Petra city in the rock we model to our terraforming needs. In the centralized core of the pyramid grid there are ocean biomes separated apart from one another. The ocean biomes host a vast array of marine life from the earth. Human colonies live around each ocean beyond a safety perimeter.

Think of a grid paper laid out on the surface of Mars surface horizontally. Each separated square on the grid is a specific dimension representing a mapped vertical volcanic chamber going deep into Mars. In some squares there are circles representing chambers in the Martian interior which will be flooded and made into living oceans, seas, and lakes. In the subterranean depths of Mars dormant chambers will host life. Martian residents will live around the outer perimeter of dormant volcanic calderas beyond the crest rim of craters for a mile outside the water's edge. Imagine the volcanic chamber on Mars being repurposed into a Martian lake that extends downward for miles becoming a vast fish farm aquarium containing marine life from the earth.

Yes, this project is a massive undertaking, but terraforming planets isn't an easy task for any engineer, scientist, inventor, or skilled worker all working together. There will be tunnels way out beyond the pyramid steps for each tier surrounding its perimeter and extending out vast distances in a variety of shapes. We'll build our pyramid utilizing what Mars naturally gives us beneath the surface from the volcanic tunnels its provided us. Martian volcanic tunnels repurposed for a homeostasis environment which is surrounding a pyramid interconnecting to the complex which is centrally located is a space mission in the right direction. The sooner we send candidates to Mars to encircle it and return back to Earth safely, the sooner society will accept repetitious missions to Mars.

Image courtesy of NASA

Universal Space Exploration (USE) has begun the seismic mapping of dormant volcanic chambers on Mars. The terraforming of Martian tunnels is important for the grid that will service the pyramid's subterranean steps descending deep into sub-levels. The planetary engineering of Mars requires that colonists intensively labor on each tier of the pyramid engineering a variety of self sustainable ecosystems. The natural intertwined topography of the dormant volcanic tunnels within the outer regions on Mars will be repurposed for establishing zoological biomes. Those Martian zoological biomes are attached to our pyramid's perimeter. The Martian biosphere tunnel circuits will surround our pyramid's centralized position providing maximized access to colonists terraforming an assortment of territories. The need for expansive Speleology, horticulture, agriculture ranches, and wildlife grazing fields beside streams or rivers is a requirement for Martian tunnels. Airlocks will be attached to the central pyramid complex to move breathable gases around from one chamber to the next using differential pressure in a sophisticated calibration. The pyramid's homeostasis will be connected to the outer volcanic tunnels homeostasis establishing a equalizing effect through the hyperbaric transfer of oxygen and carbon dioxide. The many biomes in the entire network of tunnels attached to the pyramid will be sustaining all metabolic processes for life. Colonists will have recreational activities like found on earth in their Martian underground labyrinths to overcome many difficulties with living far away from earth. Our worldwide politics must change for humanity to adapt righteously for a life beyond our earth. This generation's astronauts are ready to pilot newly designed spacecraft to another world so lets not become a hindrance to their dreams.

We need at least two to four shuttle rocket launches every year in our space administration with trained mission ready teams to venture to Mars. Saving our humanity shouldn't take a back seat to anyone especially those opposing the cost of an endeavor aiming to indefinitely provide future generations a prosperous life. Martian tunnels extend for miles around the pyramid encircling it with an assortment of vast terrains and topographies synchronized together. The tunnel's topography moves fresh water or salt water in rivers and streams engineered to achieve a desired result for a self-sufficient biome. Each tunnel ring's geometry is wide enough to encapsulate migrating wildlife grazing throughout the environment. Domesticated animals move through separated tunnels around the pyramid in they're own biome.

After one year of grazing herd animals will return to fresh knee-high wild grasses that have grown in sections of tunnel in their absence. Mars will provide the dormant volcanic tunnels with unique topographies extending up, down, or along hilly terrains into vast plains containing Earth's biological species. Grasses on Earth may grow larger on Mars, which is good for the Clydesdale stallions grazing. The rings around the pyramid are each designed for the livestock living within them. The gases that come out from each ring are regulated by the living species strategically positioned within each tunnel to complete a specific task.

Some plants are known to extract specific harmful chemicals from air and are grown internally in the ecosystem, where they're most needed for efficiency. The subterranean pyramid on Mars is somewhat like step pyramids found in Central America and South America but on a much grandeur scale for a terraforming ecosystem.

Extending out from the rings to the pyramid itself are connection tunnels also with different forms of plant life, opened periodically for cross-pollination. Some Martian tunnels help in creating a vacuum effect, which moves air within them through differential pressure. The HVAC heating, ventilation, air conditioning, and refrigeration circulation processes within the Martian tunnels are all regulated for efficient homeostasis. As carbon dioxide rises upward, plant life utilizes it to help create fresh oxygen. The chambered fresh oxygen is funnel siphoned down into vast tunnels connected to the main pyramid on each level. Oxygen is used by human beings and zoological species

on each tier and each step of the pyramid.

The zoological creatures and colonists breathe in the fresh oxygen expelling carbon dioxide which rises inside interconnected tunnels attached to the tiers of the pyramid enabling respiration processes to repeat in precise calibration. The exchange of carbon dioxide and oxygen inside the step pyramid is a natural process establishing an equilibrium of gases to move freely in hyperbaric synchronization. Extending down from each tier of the step pyramids are a series of waterfall types, each with its own unique plants, trees, ponds, lagoons, streams and river currents. Some of the waterfalls spill out to create rivers that open into a mouth, developing underground estuaries. Natural and artificial sunlight enters from the top of pyramid descending into each of the waterfall chambers reflecting light to the surrounding species.

All the rivers perform different functions, filtering pollutants and sediments before returning to the central oceans, seas, and lakes in the center of the pyramid, extending up like pillars for miles. All human, animal, and decomposable waste is processed in the lowest levels of the pyramid into compost or other usable materials keeping gases separate from respiration. Insects, worms, and other species are bred and fed to plants and other wild creatures. The gases moving throughout the pyramid ecosystem all interact with precision so that homeostasis is maintained.

Eight large waterfalls and eight smaller waterfalls extend from the top of the pyramid toward sixteen tunnels on each level for each varying shape of the surrounding rings. Freshwater streams and ponds are accessible to the livestock within each tunnel ecosystem. There are many different species of birds within the subterranean chambers. Some tunnels have bison, which are kept separate from other species of cattle and wildlife. Separating wildlife species is to prevent (CWDs) chronic wasting diseases or bovine diseases from developing or mutating within chamber.

Every effort will be made to eliminate unwanted diseases from entering the ecosystems we create, but we still have the human factor. Intentional or ill-made mistakes human beings make could create chaos in space colonies, and thus the selection process to be a colonist must be the most stringent. Also, many tunnels contain animals that the colonists do not consume, which are strictly for zoological preservation and conservationist purposes. Endangered species on Earth are bred in these underground Martian tunnels and live in habitats free from harm, other than that from other species within the maze of chambers. All the gases to sustain life in the underground habitat require an emergency threshold in the homeostasis. If one system is compromised or disrupted, becoming toxic, this threshold allows for the movement of gases to reduce and eliminate toxicity.

Emergency thresholds protect against low oxygen content so other systems balance the equation and so no species are harmed. In the center of the pyramid are two frigid ocean columns, two mildly warm ocean columns, two sea columns, and two freshwater columns, for a total of eight different biome habitats. Melted ice on Mars floods and fills the modified sealed-off volcanic chambers creating am encapsulated system for marine life to thrive.

Melted water from the Martian ice fields will be guided into a water filtration system, purifying it before dispensing it into a vast network of drip irrigation channels. The dripping springs will be directed toward horticultural gardens stretching for miles in heated germination tunnels beneath the Martian surface. Some hanging Martian crops will vertically descend from the ceiling to the floor in vast rows inside pots filled with nutrient soil providing colonists fresh produce. Martian vineyards, and wines will become highly prized vintages for future generations. Martian dripping water irrigation channels shall also feed the surrounding perimeter tunnels filled with vegetation various animal species rely on for survival.

Marine life from the earth will be inserted into the deep flooded volcanic chambers, creating new habitat within tunnels, caves, crevices, ledges, reefs, and chambers for many marine species to reproduce. Zoological species from the earth are placed within Martian tunnels to provide sanctuary breeding grounds. The ring tunnels also have areas within them that open up into vast widened chambers like open prairies for the animals to roam about free and apart.

How is a project like this built? The answer is through programmed robotic tunneling equipment through a (CNC) computer numerical control system connected to seismographs. The best-quality industrial bort diamonds from Earth will help us bore through Mars with precision. Even if humans aren't near the Mars equipment it's doing the tunneling job machining it's way following a guidance preprogrammed schematic from a internal computer numerical control system.

We will engineer the entire project for the Mars pyramid architecture here on Earth because we've mapped beneath the Martian surface with our rovers. Then our machines will do the tunneling work of creating our design with precise navigation systems connected to seismographs directing each movement beneath the surface of Mars. Robotic systems we create here on Earth will be the robots that maintain our drilling robots on Mars, creating our tunnels for terraforming.

Human beings can build cities, tunnels, railways, bridges, dams, and rockets on Earth; why can't we build a pyramid ecosystem beneath the surface of Mars? Such a habitat will enable teams of scientists to live and work together on Mars for as long as they desire.

Geologist teams can mine our prospected asteroids for valuable metals and rare usable substances for our spaceships, construction projects, and technology. Why is this idea called the Pyramid Project? Because around the earth, numerous cultures and people built pyramids.

Since past civilizations could build pyramids, so can we. We are designing them with one purpose in mind: the survival of the human race. Should a meteorite or asteroid threaten the earth, we can save human life and millions of species. If human beings destroy Earth through pollution, we can save humanity and millions of species.

If the human race ever experiences a global plague that does not have a cure, we can save humanity and millions of species from infection. If governments around the world ever decide to launch a thermonuclear war and destroy the entire earth, we can save humanity and millions of species. If natural disasters make the earth inhabitable, we will have a choice of where to go to live out our remaining years within a Martian ecosystem.

Think of the Martian outpost itself like the pillars of skyscrapers stacked together underground, forming the shape of a pyramid. Eventually, many of the luxuries we have on Earth will be accessible on Mars, with future generations all working together. A vacation to Mars and back to Earth could be in our future if we start to develop the pyramid now.

I understand many will not join the cause, but should any one of the aforementioned events take place, the investors will be triumphant, living in our Martian space ark. We can design the three-dimensional models now for Mars and begin the conversation about implementing an immediate plan. The sooner we start using our time wisely, the sooner we will leave the unbelieving in the dust when catastrophe strikes.

Yes, these plans are grandiose, and that is why we make them—

because they're hard tasks to accomplish. We must do the work others will not do because that's what it takes to set us apart from them as we build our legacy. If people don't help us, so be it, for history will show we tried before any catastrophe occurred that could have had a master contingency plan.

Global societies must labor together on a massive scale to protect our ecosystems from a cascading collapse of life and infrastructures dying. By planting trees and developing artificial reefs for numerous species, we can prevent the massive collapse of ecosystems harmed by pollution. The cross-pollination of trees, flowers, and shrubs is critical to supporting humanity's existence. Insect kingdoms that depend on vegetation for their survival are being deprived of life due to human activities.

The highways, roadways, bridges, cities, canals, and aqueducts of our modern societies were once forests or fields supporting insect kingdoms. Animals that used to thrive in their geographical habitats are now imprisoned by humanity's man-made perimeters constricting their movement. In the past, the movement of animals was never restricted except by natural mountain ranges, valleys, and waterways, which separated one species from another.

There's a problem with the way zoologists think these days because when Yahweh made everything in this world, the design was good. The designs of Yahweh are being ruined by humanity infringing on the living space of the earth's species. Humanity is not respecting the well-being of every living creature on Earth. Zoos are torturous to the animal spirit, confining them, for they were created to be free in their vast natural kingdoms.

Imprisoning wildlife within a small perimeter doesn't permit them to be wild creatures. All wildlife creatures have genetic characteristics that require them to be free, as was the natural design before humans stole their freedom. Stealing the freedom of animals is criminal, and when creatures bite back, they oppose what's wrong.

The zoologists who support cruelty by imprisoning wildlife don't really understand nature and how to properly treat animals. Humanity must develop new habitats around the world to replace the ones destroyed. Earth doesn't have a homeostasis for wildlife currently because humans aren't being fair to the needs of Yahweh's creatures because of reducing their environment. Planting trees and developing artificial reefs with materials approved by scientists is crucial to halting global warming. Nations must not contaminate or harm marine species further by placing more chemicals in the seawater.

So much human pollution has been causing irreparable damage to ecosystems around the world. Polluting needs to stop! If humanity kills the earth's species, humanity will murder the cures for diseases affecting humans; therefore, we'll be murdering ourselves.

Many medicines derive from serums synthesized from marine animals and numerous living creatures. In the Holy Bible, the Creator Yahweh said that those who destroy the earth will be destroyed. If we destroy the world carrying the cures for diseases, we will be destroying ourselves. Don't be like those who are contributors to the earth's cascading death by not laboring to halt global warming. Let's germinate billions of trees and build billions of reefs together. We need to be cleaning up our environment for the earth's species.

Earth has nearly nine billion people, so we have eighteen billion hands able to labor toward stopping global warming. Let's start a global diving program in all nations to go out and clean up the waterways, lakes, rivers, streams, oceans, and seas together. Newly equipped diving vessels properly maintained by trained divers can usher student divers to clean up our environment. There are many yachts and vessels in marinas that can host students to labor together on environmental excursions to accomplish cleanups. The before and after cleanups of the environment will be visually rewarding for students watching wildlife reclaim their habitat.

The philanthropist contributors to these diving programs and vessel greenhouses replenishing islands with new life can educate young adults vastly in the life sciences. We need more young members of our society becoming marine biologists, botanists, arborists, scientists, and oceanographers. We need to facilitate classrooms, training, equipment, and voyage excursions for diving students who can unite to make international environmental improvements.

1. Trees give us clean oxygen for all to breathe, sustaining life.
2. Trees remove carbon dioxide from the upper atmosphere.
3. Trees provide homes for eagles, hawks, and other nesting birds.
4. Trees provide habitats for wild species, such as cats, monkeys, and bats.
5. Trees provide safety for many reproducing species.
6. Trees' roots cleanse small amounts of toxins from waters.
7. Trees' leaves cleanse chemicals from many waters.
8. Trees provide forest creatures habitation for their offspring.
9. Trees provide fresh food for world civilizations to eat.
10. Trees provide medicine to heal illnesses and diseases.
11. Trees provide hammock locations on hot days for our habitat.
12. Trees provide erosion control around farms.
13. Trees provide humans wood to build homes and other products.
14. Trees provide living creatures homes in nature.
15. Trees provide sanctuary for peaceful walks.
16. Trees provide flowers for bees producing honey.
17. Trees help the cross-pollination of crops.
18. Trees provide wood for furniture, art frames, and espaliering.
19. Trees provide wood for boats and ships to carry people.
20. Trees provided us the cross Jesus Christ was crucified on.
21. Trees help shelter us from strong winds in storms.
22. Trees help combat global warming.

23. Trees help all nations prosper with reforestation resources.
24. Trees provide us raw materials to make paper products.
25. Trees give us oils for a variety of ointments and lotions.
26. Trees provide us incense to lift our prayers to heaven.
27. Trees protect us from intense sunlight, shielding our skin.
28. Trees provide us natural wild silk clothing and suture fibers.
29. Trees provide cultivar propagation experiments.
30. Trees provide atmosphere inside scientific biospheres.
31. Trees provide organic chemical compounds for pharmaceuticals.
32. Trees provide the ability for off-world colony projects.
33. Trees provided the budding rod of Aaron in the Torah.
34. Trees provided the staff of Moses in the Torah.
35. Trees provided the rods to carry the Ark of the Covenant.
36. Trees provided beautiful sanctuary in the Garden of Eden.
37. Trees provide nuts and berries for animals to forage.
38. Trees provide leaves for mulch to help other species grow.
39. Trees provide sound in the wind as a symphony of peacefulness.
40. Trees help save lives in storms, giving themselves to hold.
41. Trees provide us genetically engineered biological fruits.
42. Trees provide us charcoal to expel poisons from our bodies.
43. Trees provide us habitats to dwell among them in peace.
44. Trees provide us wooden musical instruments for symphonies.
45. Trees provide us scientific methods to measure weather.
46. Trees provide us historical records of thousands of years.
47. Trees provide us beautiful petrified stones for artistic uses.
48. Trees provide sanctuary around our deceased to rest in peace.
49. Trees provide warmth to families in winter's cold.
50. Trees provide estuaries for frogs, fish, and other species to live.
51. Trees provide spiders a means to reduce insect swarms.

Artificial reefs provide many benefits in mitigating the effects of global warming on our planet's ecosystems.

1. Reefs provide marine life habitat to create the oxygen we breathe.
2. Reefs provide ocean creatures places to reproduce.
3. Reefs provide expanding territory for marine organisms' growth.
4. Reefs provide sites for scientific research on ocean currents.
5. Reefs provide sanctuary to many endangered marine species.
6. Reefs provide a retreat from overfishing, preserving offspring.
7. Reefs give rare species breeding grounds.
8. Reefs give divers opportunities to improve the environment.
9. Reefs give divers project opportunities to clean the oceans and seas.
10. Reefs give fresh, clean food for spearfishing sports divers.
11. Reefs provide fishery science divers monitoring opportunities.
12. Reefs help protect smaller marine species.
13. Reefs help scientists monitor changing currents and temperatures.
14. Reefs help oceanographers photograph new life transformations.
15. Reefs help preserve places endangered by human pollution.
16. Reefs bring divers to waters not visited, aiding research.
17. Reefs help coral adhere and replenish on man-made structures.
18. Reefs help prevent tidal erosion close to shorelines.
19. Reefs help near-shore fishing from scientific labs.
20. Reefs help transform barren marine areas into new ecosystems.
21. Reefs help marine ecologists and biologists study life.
22. Reefs help produce microscopic organisms to feed marine life.
23. Reefs help marine species thrive for protein biomolecules.
24. Reefs work with estuaries to create interconnected habitats.
25. Reefs help kelp forests and mangrove trees flourish.
26. Reefs help encourage coral propagation, raising life fragments.
27. Reefs help marine science restoration of bleached coral reefs.
28. Reefs help benthic organisms adhere to protected areas.
29. Reefs help reduce coral disease by expanding territories.
30. Reefs designed to capture plastics help protect natural reefs.

31. Reefs bring together builders and engineers to fix problems.
32. Reefs provide near-shore fish farming to feed civilizations.
33. Reefs help researchers plan for global food needs.
34. Reefs help researchers with new pharmaceuticals.
35. Reefs help microbiologists study global warming problems.
36. Reefs help young divers explore for future research programs.
37. Reefs help establish new zones for conservation reserves.
38. Reefs help fish replication to help feed billions of humans.
39. Reefs help bring people together for nature.
40. Reefs help decomposition and processing of natural organics.
41. Reefs help remove minute amounts of toxins from marine waters.
42. Reefs provide fossil records of Earth's temperature changes.
43. Reefs provide us beautiful coral and minerals to appreciate.
44. Reefs provide opportunities for off-world colony projects.
45. Reefs help combat global warming with life adaptations.
46. Reefs help remove carbon dioxide from the upper atmosphere.
47. Reefs help reduce strong currents in storm surges and tsunamis.
48. Reefs provide us chemicals that expel toxins from our bodies.
49. Reefs provide us chemicals for restoring healthy cells.
50. Reefs help to cross-pollinate for new marine life processes.
51. Reefs provide aquatic ecosystem research for biospheres.

Coral reef ecosystems are dying because of global warming, pollution, plastics, and diseases coming into contact with marine species. The dead, bleached coral reefs affected by changing temperatures need environmentalists and scientists to alter their approach to dealing with the problem. Reef habitats for millions of marine species must be established by scientists and divers working together, creating protected zones for endangered species.

Coral reefs predominantly establish themselves in thirty to sixty feet of seawater, a no-decompression depth for recreational divers. Temperatures are rising at this depth, killing reef ecosystems. The answer is helping coral establish itself in deeper, cooler waters. Coral transplanted from shallower waters to deeper waters along descending coastal slopes is given a second chance at life. A prospective project using modern technologies harnesses solar power from panels incorporated into the exterior of buoys to light artificial deep-water reefs. Luminescent umbilicals would create artificial sunlight, generating awnings that illuminate aphotic and benthic zones. The solar buoys' strong umbilical cables would be anchored to the ocean floor.

Lower photic zones would be areas for establishing new reefs that border aphotic zones. The photic zones are easily accessible by recreational scuba divers in shallower depths. Commercial divers would work in the aphotic and benthic zones for deeper underwater projects requiring precise engineering.

In deep, light-absent waters, tethers shall extend up to the surface, collecting solar power and relaying energy to underwater awnings secured along ocean ridges above reefs. Artificial sunlight shall be reflected downward from canopies onto artificial reefs. Underwater video cameras will record all research activities at the site for a documentary.

The casing for the retractable awning is to mimic the appearance of the coral reef so as not be an environmental eyesore as it extends outward, telescoping to an overhang position. During the day, telemetry devices regulate and gather the computer-generated sunlight produced by the awnings for distribution. During the evening, the sunlight awning is retracted into its weatherproof horizontal housing for protection. Quarterly or semiannually, divers can inspect devices used in the artificial reef system to perform maintenance and ensure proper functioning. Some protected areas near the site can have underwater habitats or hotels for research divers studying the reef's ecological biome.

This artificial reef will incorporate interlocking sections of concrete, rebar, stone, and gravel assembled for surface attachments to begin electrolysis on the pyramid habitats. Sections assembled on the ocean floor in an oval will take on the shape of a colosseum that marine species may swim through on multiple levels. Research divers will monitor the site's ecology and changes, recording the fish species that visit the sanctuary.

One program every country on Earth should make mandatory for all students is free breakfast and lunch, provided at all schools. If a child, teenager, or adult is enrolled in school, these meals will be provided at no cost. The food will be provided through a globally endorsed governmental agricultural program bound to the educational system in which no one learning is left hungry or thirsty. Emphasis shall be on not wasting food but rapidly distributing it in a government-funded program.

When a farmer is considering throwing away a crop, milk, or any food type, there should be trucking companies ready to take the food to schools for distribution or to fish farms. In national emergencies, natural disasters, and pandemics, no food should be wasted. Parents can also receive ration boxes for their homes from the schools distributing freight delivered to their locations. In cities with millions of people, many families can benefit from the food farmers consider throwing away, especially during national emergencies. Proper nutrition is extremely important for growing students. Students must be provided the right amount of nutrients for their health and educational processes. America's education system should be connected to America's health-care system so no students should be hungry, or thirsty affecting both their physical and mental health needs.

Food must be free to learning students. It's a parental responsibility to support this global program to enhance agricultural, horticultural, and fish-farming activities that produce food for all schools. This food production program should be a global standard made into a worldwide law supported by the United Nations. Nutritious meals increase students' attention, thereby allowing them to be better students.

Many nations already have school uniforms. This is helpful in bringing the focus more on social and character interactions rather than giving students the means to offend or belittle others who aren't wealthy. We need better social interactions between students to reduce violence. Improving extracurricular activities is crucial to establishing upright character in students because more time will be spent on improving social interaction to reduce problematic behaviors.

Some of my goals depend on semitrucks delivering food to summer school programs when school is not in session. This way, students in larger cities continue to receive meals, and they're not going hungry or thirsty during the summer. I plan to run this program independent of the government, utilizing the Catholic Church because government offices have been in opposition to me. Hindering and stopping me from earning wealth to fulfill my goals for the past thirty years are obstacles I don't need. The future of mankind is in establishing technologically advanced programs for agriculture, horticulture, and environmentally protected fisheries while creating new biome environments in other worlds.

Where a program is established on Earth, it's not impossible to develop a similar program in another world with the right leadership. All humanity must be able to eat and drink, no matter where they reside; this should be a standard even for off-world universities or communities. Future facilities on Mars, Jupiter's moons, or in other worlds have the potential for planetary engineers to terraform internal surfaces.

Another prospective project is uniting all nationalities of people to instruct their children to start environmental projects to improve their communities. Parents and teachers must guide children to physically do the work, planting trees as soon as they begin going to elementary school. All nations must agree to reduce global warming together, planting billions of trees in a unified globally consolidated effort. Billions of people on Earth need to unite for something, and the environment affects us all.

Trees help us all, so we need to help all of them by preserving nature as our biological responsibility to enhance life in Earth's biomes. Many forms of bamboo can be used to reduce carbon dioxide in the environment. Vacant desolate islands around the world with little to no vegetation should be visited by greenhouse ships. The greenhouse ships would have numerous species of plants and trees ready to be planted to reduce global carbon dioxide levels. Ports in many locations worldwide have marina crews that could be providing a service, planting indigenous seedlings on nearby uninhabited islands. Planting trees helps reduce soil erosion. Mangrove trees are beneficial to many island habitats, providing sanctuary for many species.

Carbon dioxide released from burned trees adds to global warming. Many states, like California, need subsurface water tanks to hold fresh water for local emergencies in towering underground silos. Massive wildfires endangering communities need many convoys of electric semitrucks pulling water tankers to these emergency sites to protect neighborhoods. Cities threatened by fire need to have water towers scattered across many states, reducing the billions of dollars spent annually to combat fires. Convoys of trucks encircle a water tower, which funnels water rapidly into all the tankers simultaneously from large spigots. The tanker trucks fill to capacity in a minute. One group of trucks leaves quickly, and another group of trucks circles to refill and go combat fires with water cannons.

All the trucks are fireproof with cabs that protect the firefighters inside. The water towers serve a dual purpose as fire stations and emergency medical services stations. The water towers also have underground housing or housing built right onto them. Police divers can also train at the water towers, entering the underground water silo pools.

Drinkable water is very important during forest infernos, earthquakes, storms, floods, terrorism, or other natural disasters affecting different city systems. The water towers should be built on top of mountain plateaus with tunnels being dug to erect the underground water silos beneath the vertical towers. High towers are also great observation posts and launch platforms for drones that can gather various readings for gases and wind currents, collect video, and improve communication and signaling.

Water tankers can be designed for multiple purposes or different trailer attachments, depending on the need. Electric semitrucks can aid in removing debris, delivering drinkable water, carrying supplies, and even moving people in emergencies with some tanks to supply breathable air. The combination air/water cannons on the trucks can shoot fire-retardant materials that can strike distant trees, buildings, and surfaces, slowing or stopping the progress of fires.

Stopping Americans from being harmed by fires and other events is important to the security of this nation. Planning ahead and building many of these towers as possible across the United States will be beneficial, as lower coastal areas are endangered by hurricanes, rising ocean levels, tsunamis, and torrential storms. Elevated water towers on mountains reduce the chance for water contamination, which can occur at lower elevations vulnerable to aquifer contamination after flooding.

Fleets of electric trucks with hybrid capabilities provide longevity to aid in many emergency scenarios where the need arises. NEST can be a partner in these water towers and underground water silos, which can have nuclear fallout shelter condos in their basements. This plan maximizes the multiple uses for structures, justifying the construction costs associated with the undertaking. These projects will not just be single-use buildings erected with state or federal budgets from government resources being not utilized.

These buildings will help in numerous emergencies differently than FEMA because they will be in the control of the states. Instead of states waiting in emergencies for the federal government to make a decision, these water tower facilities will be immediate service centers to aid the public. The National Emergency Strategic Task Force and the National Emergency Security Task Force operate two distinct training regiments. NEST members in each state help improve the recovery time from natural disasters to aid Americans promptly.

Danger from tornadoes, hurricanes, floods, earthquakes, fires, and volcanic eruptions has become part of American life. Danger from terrorism, mass shootings, home invasions, identity theft, homelessness, and numerous other crimes has also become part of American life. Sometimes, when one disaster occurs, this, is turn, creates many more emergencies as some Americans resort to criminality. With millions of Americans who have committed crimes, the dangers in society are more apparent, requiring good citizens to have a militia ideology to protect their families and friends. NEST members are emergency workers and militias separate from the National Guard and designed to care for the needs of communities in a religious spirit.

The general public can apply for NEST membership, much like in a fraternity or sorority, but it comes down to being selected. NEST members honor a set of values, believing in being fruitful in all holy works that lead to sainthood.

The quantity of nuclear bombs detonated throughout the history of the world may have been a contributing factor that's caused coral reefs to die. Scientists say it's global warming that's killing the fragile reef ecosystems, but bombs can affect a coral reef system in a physical and chemical reaction.

A damaged upper atmosphere permits more radioactive ultraviolet sunlight in the earth's aquatic environment. These nuclear reactions do generate heat within the upper atmosphere in addition to increased radiation. It's very possible that all the nuclear bomb testing that was done in the Pacific and other areas around the world sent shock waves through reefs, causing a chemical ecological reaction.

The heat generated by a nuclear bomb near a coral reef is high enough to change surrounding water temperatures. A thermonuclear reaction does generate large quantities of heat, and when hundreds of tests are performed on islands with reefs, a chain reaction can occur, affecting life. We know radiation can be absorbed into living organisms and remain there for decades. Global warming is real, but it can be secondary to the initial injuries done to coral reef ecosystems.

The energy transference from bombs has compounded temperature changes that have affected the electrons in coral reefs. It's a culmination of many factors affecting or harming coral reef ecosystems, not just fossil fuels, pollution, radiation, chemical reactions, heat, and sound. Nuclear and other types of bomb testing have sent shock waves through sensitive ecosystems, disrupting them. This is part of a hypothesis regarding the bleaching of coral reefs throughout the world.

Governments are spending billions of dollars to destroy each other rather than spending resources on saving our planet from extinction. The earth is dying and not from any other species but human beings. How can nations look at themselves in the mirror knowing that they're contributors to the earth's destruction? Nations steal from one another and try to dominate one another, but all the energy put into thwarting each other's efforts has destroyed our planet. With the world's population rising, nations that lack education in sanitation, recycling, agriculture, and health care need to be elevated to reduce the risk of worldwide pandemics.

Human waste in all its forms affects our environment, recirculating back to human beings through exposures or absorption. Poor sanitation leads to more diseases emerging, which has increased mortality in third-world countries. Powerful nations must make it their focus to aid nations that lack strong educational systems. The sharing of art materials aids a student's creativity, allowing them to have a positive impact on their environment. Many countries are utilizing recyclables to erect permanent housing, and this was all derived from art students creating new designs.

Space tourism companies seeking to explore beyond Earth have a crucial importance to saving living organisms from being destroyed by mankind. We already know too many nations have nuclear, biological, and chemical weapon systems that can destroy planet Earth. For thousands of years, nations have fought, killing one another, including today. To be wise is to plan ahead, protecting what we already know mankind will destroy. Nations must accept that human beings have a destructive history. We must figure out how to prevent environmentally identical scenarios from happening in other worlds.

Preventing damage to the planets we will flee to for survival is important. Human beings have destroyed coral reefs, polluted rivers, poisoned air quality, and poisoned our oceans with numerous chemicals. Natural disasters like hurricanes and floods moving entire cities' trash and sewage out to sea is disastrous to living organisms trying to survive. These global catastrophes are disturbing to me as I pursue becoming a planetary engineer to aid life.

Universal Space Exploration (USE) is a new research company aiming to launch spacecraft, conduct space tourism, and colonize other worlds using new technologies. Mankind's survival resides upon preserving the Earth's species. Already pollution, global warming, and environmental disasters are three contributors that harm sensitive ecosystems. Developing a space ark to transport Earth's life off world is one method to protect already endangered life. Other planets in our solar system should be used for storage facilities, containing vast amounts of life from Earth. I'm not just talking about seed banks on other planets but also zoology facilities for the preservation of many living species.

Universal Space Exploration has begun the seismic mapping of dormant volcanic chambers on Mars. The energy grid that serves the volcanic tunnels on Mars must be installed in such a way that every system has backup power. Since the volcanic tunnels on Mars are mapped beneath the surface our marking grid is laid out over the top surface. Our surface grid marks the positions of tunnels below the surface this aids our pyramids placement for a central position using the volcanic chambers in the most efficient way. Complex designs can be accurately bored when we know exactly where the tunnels reside in our interconnecting process.

Each descending step of the pyramid being built in the sub-levels is creating a terraformed environment. The planetary engineering of Mars requires that colonists intensively labor on each tier of the pyramid engineering a variety of self sustainable ecosystems. The natural subterranean topography of underground tunnels on Mars, in the outer regions beyond the centralized pyramid will be repurposed to establish zoological biomes and agriculture. Around the pyramid's perimeter airlocks will be interconnected intermittently to equalize living environments for homeostasis. In the Martian underground labyrinth colonists will have an assortment of activities to help them cope with the difficulty of living so far away from earth. This generation's astronauts are ready to pilot spacecraft to other worlds and Universal Space Exploration desires to speedup the process.

On Earth, numerous species forage, traveling in herds, packs, swarms, schools, colonies, and groups. On Mars, volcanic tubes, including the excavated chambers we erect to create subterranean labyrinths, can be designed to permit two-year migration times for each species. Species transferred from Earth to Mars will travel along tunnels, consuming new growth along horizontal wide plains migrating corridors in waves, grazing on new life. Landscapes eaten within the man-made Martian tunnels will regenerate new growth as the species make their revolving journey, traveling expansive distances. If the tunnels are made in subterranean colonies around the entire circumference of Mars, then mass production of agriculture can feed increasing populations.

Vegetation can fill tunnels on many different levels like stacked, hundred-mile-long doughnuts holding agricultural and zoological species in an ark of life. Species like cattle, sheep, deer, bison, buffalo, alpacas, llamas, chicken, turkeys, emu, ostriches, pheasants, cows, addaxes, antelope, gazelles, giraffes, goats, ibex, geese, quail, and doves can live in the Martian tunnels. Animals must also exercise to maintain their health.

Mining in space must focus on construction: for example, finding ways to create concrete from minerals we have access to on the red planet. Once we perfect Martian cement we can use the sprayed concrete hardening material onto vast tunnel labyrinths. Volcanic chambers can be enlarged and modified to accommodate highway-long tunnels of grazing fields for agricultural purposes. The United States has numerous highways stretching the entire country from east to west and north to south. We should think of massive cylindrical tunnels as wide as eight lanes stretching for many miles in a doughnut configuration being a sacred torus shape beneath the Martian surface. In America, we already have mile-long tunnels deep underground. Enlarging volcanic chambers on Mars won't be as difficult versus dealing with all the hardened stone like we have here on Earth.

Huge tunnels on Mars, very wide and very long, can be flooded with schools of fish that can swim in many directions through flowing underground rivers in labyrinths. The walls inside the volcanic tubes on Mars can be bioengineered to look like aquatic biomes here on Earth. As seen in massive aquariums, these fish colonies can be replicated underground in enclosed chambers on Mars. Fisheries can breed schools of fish for the human colonies living on Mars, and the inhabitants can scuba dive in the aquatic biomes performing science diving.

Mining for metal should also be a priority to aid construction on Mars as well as in other worlds nearby. Companies outer space mining for valuable resources on planets, moons, and asteroids have new techniques, methods, or technologies ready to deploy. The tunnels on Mars shouldn't just be circular but should be almost like the cave-diving chambers found on Earth, but more massive. We want the fish on Mars to find their own protective habitats within the labyrinth of tunnels along the planet's subterranean circumference. The underground chambers can be separated altogether or periodically opened for short intervals for such needs as feeding.

The underground caverns beneath the Noctis Labyrinthus region will be closely examined, as will other locations near Mars's ice fields. The Mars mission should immediately begin its operation mining for resources such as water, agricultural soil, and volcanic chambers for housing. Teams shall select the most advantageous communications sites on Mars to enhance their missions' success and relay vital information back to Earth, maximizing every tangent of efficiency. Should one communication site be disrupted by storms, others will be able to transmit.

Horizontal and vertical chambers on Mars will have modifications so colonies can have a space that can provides them a comfortable, sustained living arrangement. Loft-style bunk beds or cave cubicles shall be established in the walls of the chambers from the floor up to the ceiling. If tunnels are engineered or volcanic tubes are modified to have high arc ceilings, then these chambers can be used for housing while filtering out gases. Attaching platforms to walls or carving out cubbyhole rooms in increments up walls can provide thousands of homestyle habitats that are like miniaturized cave apartments or condos for colonists.

Like a cave on a mountainside, there can be thousands of spaces mined out for tent chambers within long tunnels with high ceilings. Each side of the tunnels can have residences with ladders attached between them for entering and exiting bunks. Also, there can be stairs to the chambers carved in the walls of the tunnel.

Mars may have three or four different types of living arrangements for colonists. New arrivals may receive a walled bunk bed until they do the necessary work to dig out their chamber in another section of the complex after a long day of work on their survival duties. Digging out chambers will be something new arrivals can do in their off-duty time.

The sooner new arrivals do the work to dig out their chambers, the sooner they can move out of the military barrack area and into a new style of living with more privacy. Chamber homes on Mars maybe be like living in Native American cliff dwellings on Earth in the beginning developments on the red planet. Tunnels on Mars can have cliff dwellings on both sides of the volcanic chambers accommodating thousands of people like a Guyaju, Matmata, Bamiyan, or Vardzia complex.

Here on Earth, platform loft tent beds will be attached to walls with ladders between the vertical tiers. These tiered living platforms in the tunnels will be for training the next group of colonists to live in confined space in fallout shelters, which are elaborate military bunker complexes. Living on Mars will be a confined until the first ten teams establish a stronger infrastructure for more workers to arrive. From the floor to the ceiling, horizontal wall-mounted tent loft beds can provide thousands of habitats in a single tunnel chamber. The ledge beds will be welded and bolted to the wall alongside ladders reaching each terrace.

Unlike camping on the rocks on the side of a mountain, these tent beds will have solid metal frames to support the mattresses. Around each platform will be a thick tent skirting like a yurt, providing an enclosed perimeter around each mattress for privacy.

We must plan for a new real estate market on Mars because residents will want home upgrades, finer dining, more spas, better movie theaters, and more recreational activities.

The colonies' positions in relation to weather must be emphasized to ensure the mission's overall success during incremental building improvements on Mars. There's also a need for security on Mars for many different possible scenarios.

Martian tunnels can have watering holes, rivers, streams, and lakes to accommodate many animal species, including the colonists. Water funneled into a reengineered volcanic tunnel could become a river, moving fluids from higher chambers down to lower chambers via inclined planes. Looping the flow of water through long, wide tunnels can create grasslands and oases for animals and livestock inside the chambers.

It is crucial for us to start engineering tunnels on Earth to perfect the processes we can use on Mars. We can use explosives on Mars to map out the underground volcanic chambers using seismic technologies operated by robotic rovers. Then we will need a game plan for which areas of Mars to build on first that have access to water. We perfect the processes of terraforming here first, and then we transfer the techniques for planetary engineering to Mars.

There are areas of the United States and Canada open for development. The Pyramid Project needs an inland location to build underground, and I think New Mexico would be right. It is separate from other states because of the Roswell incident and separate from the inland sea project, which is in California and Arizona. Also New Mexico shares borders with Texas and Colorado, which can have participants in a subterranean facility mimicking a Mars underground colony project.

Once the science is perfected, then we'll know exactly how long it will take to complete the project underground on Mars and what to do as well as what not to do. Underground biosphere terraforming isn't easy, especially while engineering rivers to descend top to bottom throughout an elaborate tunneling system. River waters that are moving must recirculate oxygenated water.

Cascading waterfalls extending out and downward around the pyramid on each tier mimicking Earth's pools, where plant life, fish, and organisms thrive. The horizontal chambers on Mars can be flooded on some levels to create cave systems with flowing currents for aquatic species to venture out from the central oceans of the pyramid's core. The core of the underground pyramid will have vertical oceans, seas, and lakes with life.

On Earth's underground pyramid facility for research we'll have a dual purpose for when a nuclear war eventually happens; the subterranean ark of life will be a safe haven from harm. A second underground pyramid facility could be in Alaska, and a third underground pyramid facility could be in Canada. If only one pyramid facility survives the nuclear war, the ark will bring back life to Earth. Maybe only one pyramid facility gets revealed to the public, and the other two are covert. The public underground pyramid facility could be a biosphere that could be visited by students learning about the space project.

When a colony lives underground for a very long time, they need activities to stay busy and have recreation. Colonists will need more than musical instruments, board games, table games, books, movies, and general hobbies to keep themselves occupied. Yes, colonists will have work, but they must also have different kinds of work.

Running a colony will require daily work duties like ranching, biosphere maintenance, horticulture, meal preparation, new construction, and keeping an everyday planned work list. All the tasks of the colony must be completed before colonists have any leisure time or fun. Underground leisure time activities shall include: swimming, sports, scuba diving, horseback riding, equestrian events, bowling, basketball, hockey, boxing events, dancing, roller skating, skateboarding, gymnastics, gymnasium activities, workout fitness gyms, and pit freestyle trials off-road BMX.

Other popular activities for colonists will be indoor hiking through the preservation tunnels and canoeing. Also, there will be downhill skiing, tubing, and snowboarding on multiple slopes down the pyramid's underground exterior between cascading waterfalls. On the surface of Mars, there will also be kite flying, rolling sail karts, kiteboarding, and other activities.

Living on Mars must have all these things. We will replicate Mars living underground here on Earth with all these activities to survive a nuclear winter above, living for decades beneath deadly radiation. Additional sites for the bio-pyramid need not necessarily be the United States or Canada but can be built in any nation desiring to survive a nuclear war that other nations started. Weapons of mass destruction are so stupid.

What about transforming sea salt into a new material we can use? We must figure out how to make strong building blocks out of materials in space. Outer space materials must give us hard, solid substances that are formidable, giving us the greatest longevity for our construction. Here on Earth, we must experiment with what we have readily available. What about sea salt made into building blocks by mixing it with cement or other compounds, then attempting to fire it into a brick? Can clay mixed with sea salt be combined with something else and made into a salt-glass brick?

What about hard plastic windmills powered by other windmills siphon suctioning seawater through ceramic pumps funneling fluids to the desert and surrounding mountains? The seawater is filtered via absorption into the ground, passing it through many layers of sediment and leaving fresh water in the dry encapsulated underground aquifer for extraction. On the surface, the land is sectioned off, for the process of separating salt from the seawater. Soil samples are taken periodically for analysis to calculate the efficiency of the filtration process. Scientists and engineers determine how many times seawater can be sediment filtered before positive results diminish in the land quadrants used for filtering. Using the earth to naturally filter seawater to remove the salt is a utility process some deserts can experiment with. All regions around the world need fresh water for consumption and finding a better way to use seawater in a conversion process is what we need.

Surface work crews operate precise fracking machinery, remote-controlled vehicles, coring equipment, and siphoning systems enabling deep lakes of freshwater to be recovered through subterranean construction designs. Water collected underground is funneled to the surrounding orchards, livestock, and farm grazing land communities.

All the water is repeatedly tested before leaving the facilities to ensure no salt is entering the pumping stations moving water to the surrounding lands and territories. Planetary engineering must begin by perfecting new processes here on Earth, obtaining environmental protection data for engineering ecosystems. All the worldwide problems affecting our environment, including poor industrial waste management, must be redesigned to prevent pollution.

We must not take ineffective processes established on Earth to other worlds, thereby polluting those planets and moons. Planetary engineering starts here on Earth but finally perfects the processes once they're implemented in outer space. Our universe needs planetary engineers who care and want to save human life on a planetary scale.

Adapting to extraterrestrial worlds requires us to master what we've learned here. Water filtration processes must first be tested in large, graduated cylinders, filtering seawater through numerous layers repeatedly. Core samples will be extracted from the project sites for analysis.

Clean water will be encapsulated in deep subterranean lakes. When tests come back, we will know if further desalination processes and water treatment will be needed. On the land's surface, if the salt content becomes too toxic, we can scrape it off with bulldozers and bank the salt in one area. Ancient seabeds that were drained long ago can be remade into new seas. Explosives can be used precisely to create crater spaces, changing a region's territory. New ecological species can be implanted into the seas like kelp, sea grasses, and new coral for marine ecosystems.

Charcoal from burned forests can be utilized to help filter salt water or add to the sea's soil composition. States like California, Arizona, New Mexico, and Texas can terraform their regions, even along the border with Mexico. Wildfires burned forests, and new forests need fresh water for billions of new seedlings, cacti, ornamental grasses, flowers, and aquatic oxygen-generating species.

Terraforming a new man-made sea also means developing marinas and housing around the lakes. The food for future generations must be started now so it's established then. Indian reservations and the desert territories need a major overhaul, becoming more green once they have access to fresh water. The silos scattered over vast regions will also provide water to fight forest fires, stop arson, and aid communities in emergencies. Xeriscaping must also occur on the mountains, blanketing them with flowers, shrubs, hardy ornamental grasses, and drought-resistant trees.

Having grandiose ideas isn't something absurd. Look at the pyramids of Egypt, the Great Wall of China, the moon landing, rovers on Mars, satellites to Andromeda, Burj Khalifa, the Hoover Dam, cross-country trains and roadways, et cetera, so we shouldn't dismiss large-scale projects, because they're not impossible. The harder we work, the more we accomplish if we focus.

Recreating an inland sea from the Cretaceous period is another project involving tunneling to siphon ocean water back into mountain ranges, making three miniature seas with three heights and depths. The three mountains' tier levels will recirculate the seawater back to the ocean through a series of rivers and streams breeding fish from hatcheries. This planetary engineering project is replenishing fish stocks and doing research for terraforming other worlds.

Windmill sprinklers will help in irrigating arid land with fresh water gathered by condensing vapor from the ocean water in the canyons of the three man-made inland seas. Replicating the earth's biomes on other planets will ensure the survival of humanity should the earth ever be endangered by another massive asteroid striking it. Dinosaur extinction is one fact to inspire us to save our world through preservation vaults scattered throughout our solar system.

Weapons of mass destruction that are studied by experts like myself point to a horrible future for all humanity, which I hope can be averted. The fact is that human beings have been murdering each other for many thousands of years without a desire to stop. My goal is to plan ahead for the inevitable because I know the truth that many will destroy this beautiful world called Earth.

Humanity is already destroying this world through global warming, pollution, deforestation, weapons of mass destruction, contamination, and extinction of species great and small. I'm not naive enough to think that the destruction of the earth is going to stop miraculously. Those who preserve the earth are caretakers of the garden of Eden, which is this entire planet. This entire planet was created by Yahweh; that's the garden of life we've been entrusted with.

We must start space missions with the healthiest candidates—people with the utmost sincerity. Don't just accept the credentials people claim to have; test them over and over again to validate they're being truthful, sincerely caring, and not being deceptive. During an outer space mission, all candidates must exercise regularly for the duration to maintain their health.

Space travelers venturing to Mars can exercise within a fluid-filled sacred torus cylindrical doughnut chamber that provides a resistance swimming environment in microgravity. Inside the pressurized space chamber, workout regimens can allow four or more swimmers to exercise simultaneously. Each space diver team comprised of astronauts, cosmonauts, aquanauts, taikonauts, and Apocanauts will train for their mission, exercising inside a fluid-filled hyperbaric swimming chamber. Apocanauts are a scientific religious group trained for outer space exploration possessing an ability to observe signs within our universe leading up to the great Apocalypse. Interpreting what others cannot when visiting other worlds is an invaluable skill.

Religious believers who are apocanauts will possess skills others who aren't devoted to Yahweh will not possess. These religious believers will have the ability to decipher ancient writing, tablets, or glyphs to recognize dangers that are extraterrestrial. Apocanauts would have a religious order. For example, some of these apocanauts may have additional abilities like telepathy, telekinesis, prophetic visions, unique spiritual gifts, and the ability to speak in different languages when filled with the Holy Spirit.

Space divers train inside a sacred torus jet-propulsion pool geometrically designed for swimming against dynamic flow rates of resistance fluids circulating around inside a cylindrical chamber. This device is called the Space Exercising Research Aerobic Physiology Hyperbaric Systems Medicine Life Sciences Laboratory Fluidity Module (SERAPHSMLSLFM or SERAPH).

The SERAPH health-care system is a device to aid outer space life support.

The SERAPH outer space module has a magnetic resonance imaging platform built into its walls, giving it versatility, and it can be operated on a stationary planetary surface like Mars. The SERAPH has many systems, all built within one multifunctional device. The SERAPH was designed for ease of use in space medicine and a rapid deployable assembly. Inside the SERAPH, there's a medical quarantine, a hyperbaric chamber, MRI equipment, a tissue immersion bio-engineering platform, and a regenerative healing chamber.

Inside the SERAPH, there's a diverse chamber for outer space: bioengineering, physical therapy, water births, emergency hazardous waste extraction, biopsy sampling, hazmat ejection port, and an aerobic or anaerobic exercising physiology chamber. The SERAPH was designed with versatility in mind for confined space within a spacecraft. The SERAPH unit was also designed for use in other worlds with gravities different than our own here on Earth. Each exercising space traveler lies down inside a tubular chamber in a formfitting, skintight hazmat diver's suit with a transparent helmet.

Once the diver's inside the sealed entry compartment, it's fluid flooded equalizing the outer pressure with the inner fluid pressure in the volume tank. Space divers are transferred through a wet lock into the SERAPH sacred torus. To maximize ease of movement in the chamber, no scuba tanks are worn except a small emergency breathing canister on the back pocket. There's a breathing hose extending from the helmet to a specially designed flexible cable for airflow. When the sacred torus chamber's fluids are rotating inside generating resistance current space divers swim against this dynamic resistance experiencing a workout. The breathing hose is designed so the least resistance is experienced within the chamber making the diver feel completely free in microgravity feeling forces push against them. Swimming in space is much different than swimming on earth. Fluids traveling fast at high speed in a pressurized doughnut tube is like being drenched by a fireman's hose close up which can hurt or feel very restrictive. Since microgravity lacks the forces we're accustomed to on earth we have to generate resisting forces that oppose us so we can stay healthy.

This way, the diving swimmer is streamlined, with liquids moving easily around the diver during their exercise session while striking them forcefully head on. The breathing hose is attached to a central harness lining the suit's back and extending up to the diving helmet. The diving suit is one piece, closed all the way up the back to the transparent helmet. The helmet has a drop-down shield for virtual reality usage to enable visual adventures while exercising.

Hard helmet diving regulators will be specially redesigned for this space chamber to aid respiration. SERAPH systems can be delivered to moons and other planetary surfaces for use by colonies use during their missions. Missions that terraform other worlds will need space divers exercising to keep fit, reduce atrophy or bone loss, and have recreational dives to keep morale up.

Divers will exercise hard, swimming in space to improve their bodies' circulation and muscular development. After swimming workouts, astronauts' eyesight will improve due to the trauma of the extreme gravity felt during rocket launches. Swimming in microgravity will help the human body to move fluids throughout the capillary-venous micromembranes, aiding proper circulation to support healing. Swimming in microgravity will help speed up the healing of injuries due to damaged internal tissues. We must protect our colonists' vision through a new healing process after these rocket launches and during their sustained microgravity stay on long missions.

A constant recirculating flow of fluids presses against our space divers, creating a physical resistance and making swimming difficult. Swimming divers experience neutral buoyancy and microgravity in the SERAPH's sacred torus chamber, but when the fluids whip around in rapid revolutions, resistance can be felt. All the fluids inside the sacred torus chamber loop effortlessly back around with the least obstructive resistance in fluid hydrodynamics.

The SERAPH recirculates the fluids indefinitely inside the sacred torus while the spacecraft is doing its yawing, pitching, and rolling maneuverability. Fluids move only in one direction at a time, according to the design configurations. A two-chamber twin sacred torus system may generate more stability by having fluids traveling in opposite directions during certain types of maneuvers. Enhancing the way we stay fit physically, psychologically, and emotionally during long space missions is crucial.

Inside the chamber, water jets out from one end and siphons through the hollow cylinder walls back around toward the front, arriving at its original position after one revolution. To create the least resistance, the sacred torus has an elliptical or circular geometry. The fluid dynamics within the sacred torus are designed to maximize currents looping around in mass flow rates. The hyperbaric sacred torus has a multi-wall chamber, each uniquely different by design. Fluid jet-propulsion systems provide a variety of intervals that create resistance intensity on many levels for space divers to swim through. Momentum fluid dynamics affects the flow velocity, density, temperature, and pressure variance inside the sacred torus hyperbaric chamber loop.

To improve the hydrodynamic stability inside the hyperbaric chamber, various calculations must be made in analytical geometrically to increase flow rate. The fluids inside the flooded SERAPH hyperbaric chamber do not necessarily have to be water. Fluids inside the SERAPH can be a transparent mineral oil or other fluid suitable for the chamber's operations. Divers in the hyperbaric chamber can see motion picture images through their clear transparent helmets or through a virtual reality headset visor. The impenetrable interior walls inside the sacred torus are all made of television screens, creating a 360-degree imaging platform. Inside the SERAPH, transparent walls protect the technologies encased behind them which are fluid proof.

On Earth, the tunnels inside marine aquatic centers show the activity of life around visitors. Inside the sacred torus, there's a similar type of visual experience but with recorded images. Images of underwater marine life flourishing throughout the oceans, seas, lakes, and submerged caves around the earth are screened to space divers. Also, marine life from the Arctic and Antarctic regions will be screened to our exercising space divers inside the SERAPH.

The self-contained underwater breathing apparatus the space divers will use will enable them to breathe freely and have adequate oxygen for respiration. Biochemical metabolic processes will be monitored inside the hyperbaric chambers from the suit worn by each diver, measuring respiratory and ventilation rates.

Helmets may have Sodasorb inserts to help reduce CO_2 concentrations from heavy respiration during fitness endurance exercises. A special closed-circuit breathing system, very compact and streamlined, will be built into the diving suit. While exercising, space divers can interact with one another utilizing the virtual reality or augmented reality system. Free-flowing air can come from either the umbilical hose to the diver or the compact built-in breathing system.

The goal is to provide all the respiration requirements for the space diver and to create the least amount of obstruction for the divers inside the SERAPH chamber. Rather than just running in space on treadmills, the crews of these space missions will be swimming virtually inside a neutral buoyancy laboratory. The space administration did not create or commission diving module hyperbaric chambers for the International Space Station (ISS). This generation of space missions should have diving chambers available for outer space, lunar landings, and planetary explorations that include Mars.

The speed of the fluid propulsion can be increased or decreased from a digital panel so the divers can be given different degrees of difficulty for their swimming workouts. A space health-care system is necessary for an ISS space hotel, lunar habitats, spacecraft, and a Mars colonizing mission. SERAPH exercising chambers are necessary for space crews as invaluable medical equipment that will enhance the longevity of scientific missions and the health of the candidates. Also, the diving suits worn by the space divers can be designed to resist all normal movements people make. This way, the space divers have a double dose of resistance, helping reduce atrophy. The patent claims for these designs have already been written. Once the developer has the schematics completed, then engineers will have to assemble the prototype.

Space wardrobe worn outside the SERAPH can also be of the resistance design so that resistance occurs for every movement a human body can make. In the bodysuits, astronauts will be in a constant workout during their entire mission except when they're resting or asleep.

Exercising, divers are able to view scenic underwater videos around themselves inside the SERAPH chamber. The scenic videos shown inside the SERAPH to stimulate eye movement are used to enhance eye circulation for astronauts. The full spectrum of light experienced on Earth, which the human eye needs, is identically replicated inside the SERAPH virtual reality system. All these visual experiences are to reduce vision loss during long missions in microgravity. The SERAPH chamber can deflect and detect radiation.

Very strong g-forces experienced when exiting the earth's gravitational pull have damaged the anatomy of the human eye. Space microgravity has reduced the flow of bodily fluids needed to repair human eye injuries quickly. There's a need to heal the entire content of an astronaut's eyes while in outer space and not leave the injured patient unattended to. Swimming in a microgravity environment improves the healing of the human eye. The more the human body pumps blood around in microgravity, the better it is for the eyes constantly needing the recirculation. Human eyes constantly moving around being in use in microgravity seeing different spectrums of light like we have on earth are less likely to have a loss of vision. The eyes need a workout just like the rest of the human body needing physical activity to stay healthy.

Two multi-layer sacred torus geometrical configurations move fluids inside the walls of the hyperbaric space diving swimming pool propulsion chamber system. One chamber moves fluids to the left the other chamber moves fluids to the right, and the result is equalization for a balanced equilibrium. I believe hearing loss is due to differential pressure like when we travel high in the mountains and then descend to deep valleys. In fetuses, infants, and small children the inner ear may not be able to compensate pressure as easily as adults. The result is that damage can occur in the inner ear throwing off the way fluids and surrounding pressures decipherer vibrations. The human body adapts to the exposures experienced so a fetuses inner ear can be disturbed by altitude pressures or hyperbaric pressures just enough to lose sensation. Loss of inner ear sensation can lead to loss of hearing altogether. Human beings often want to push ourselves to the extreme but for a fetus these extremes they aren't ready for especially in early development. Adults can handle a firecracker sound for example if it's not too close. A fetus in early development may hear a fire cracker sound like right next to their ear just when experiencing a ascent in an airplane, descending in a deep swim, or descending from a mountain to a low terrain below sea level. Pressure has different effects on the human body and a fetus shouldn't experience too much traumatic variations without being effected somehow by the experience. I came up with this inner ear theory feeling a change in cabin pressure in a semi truck cab going from high mountains to low ground quickly so I felt a dulled sensation.

In the spacecraft, the cylindrical shape can accommodate the doughnut design looping around the central habitat, or the unit can be a tubular design like an international space station module inside or attached to a spacecraft. The best designs must be incorporated into the design for the most efficient laboratory to come forth.

Three-dimensional scale models of the module must be made, considering the walls inside the hyperbaric chamber to be digital screens with a 360-degree vantage point. The viewing of screens will put space divers in a realistic new world environment in their confined space. Virtual reality face masks that raise and lower on the transparent helmets are another option, but these must cover the entire helmet to provide a real experience. No expense should be spared on this SERAPH hyperbaric chamber, which should be fitted with a magnetic resonance imaging (MRI) machine. The SERAPH module has a full medical laboratory with many features for numerous medical emergencies and services while in space. Missions to Mars or to the moons of Jupiter are expeditions that have health risks. If a colonist is injured, ill, infected, or has any type of health problem, aerospace medical doctors can attend to them with the SERAPH system.

Astronaut colonists millions of miles away from Earth will need X-rays and MRIs if they are injured. Twenty-six-month launch windows for an Earth-to-Mars factorization for any medical team is always something to remember. It will take many months or years for anyone injured to return to Earth. Reentry into Earth's gravity is increasingly more dangerous for an injured space crew member than for healthy astronauts. The SERAPH life support module is a necessary investment to save lives or even heal creatures or isolate extraterrestrial life.

Space divers inside the SERAPH experience an array of scenic environments to help them prepare for their missions. Space crews are to dive beneath lakes, rivers, and oceans of other worlds. Remote-operated video (ROV) watched in the SERAPH will be for mission training until space teams reach the final destination. ROV images can be from diving drones sent prior to the mission for digital reconnaissance footage.

The digital screens built into the walls around the divers in the exercise chamber have rotational videos of all of Earth's scenic diving locations as a world video library.

Astronaut teams who leave Earth to colonize Mars should have an extensive video library to provide themselves rewarding educational and recreational experiences. Space divers can be stimulated with different environments to swim through, making their workout experience an adventure. I hope the use of the SERAPH system can make its debut in a movie on space to show this generation what it will be like for them in the near future to go to Mars, Jupiter's moons, or beyond the outer rim.

Since the colonist crews of these space missions must stay physically fit every day before setting foot in another world, we must provide them comfort. No nation or private corporation wants their space missions to be failures, so providing some entertainment is psychologically rewarding to reduce stress among crews. A successful space mission must factor in many variables, including morale, to keep everyone happy and not make it hard, which affects minds differently.

Space divers can swim among coral reefs, shipwrecks, and under ice sheets into Antarctic reefs teeming with life. Space divers can swim to deep ocean volcanic vents and places of geothermal activity where life exists in their scenic videos. This is one way that exercise time is never boring during a space mission but always an adventure when visiting a new place.

Since the exercise chambers simulate a resistance environment, the videos can also provide space diving–related emergencies exiting spacecraft. Space diving from high altitudes in the exosphere, thermosphere, and mesosphere can be simulated in the videos. Emergency egress procedures for spacewalks turning into space diving can also be simulated in the scenic videos.

The videos used in the exercise chamber can be interactive, preparing the space team for the unexpected while refreshing their minds on the space mission. The helmets all provide sound, having an intercom voice system. Space divers can answer questions intermittently revealed throughout the video to sharpen their skills. The scenic videos can also be tests in life sciences, including an array of different subjects relating to the mission parameters.

The exploratory mapping of volcanic lava tubes making these three-dimensional videos available to colonists will better prepare them for their missions to terraform subterranean spaces. Mars and moon terrain surfaces can be made into instructional tools to enhance colonists' recognition skills for their arrival into a new environment.

Preparing the space team completely and having their minds sharp on the tasks they must accomplish will ensure the success of the mission. The videos can also be construction tasks in each phase, mimicked on Earth but now needing to be replicated on Mars as new development.

Colonists in space need guidance even while millions of miles from Earth, where minds can wander, experiencing psychological, emotional, and even medical changes. Space crew colonists need to be healthy physically, mentally, socially, emotionally, and spiritually for a favorable mission without problems. On other planets, high altitudes can also create different types of sickness affecting tissue physiology; perfect health helps reduce potential dangers for mountaineering colonists.

As the next generation of students study, they can look forward to new attractions developing for entertainment. One such improvement in entertainment will be the Gravity Planetarium amusement ride. Families will be able to experience many of the same videos, computer-generated imagery, and virtual reality that the astronauts experience at the Gravity Planetarium. Riders suited for extravehicular activity are jolted, launched, misted, and inverted inside the spherical planetarium on a film journey throughout the universe.

Like traveling futuristic astronauts on scientific explorations, riders have unlimited astronomical adventures across galaxies. Once a rider is seated, the space suit closes downward, holding them in place. Planetarium riders perform (EVAs) extra-vehicular activity spacewalks on asteroid analysis, black hole research, space planetary diving, space scuba diving, nebula observations, planetary topography explorations, and comet tracking in multimedia film adventures.

The space helmets in the amusement park ride can also provide computer-generated virtual reality experiences in the planetarium. Students may be able to sync up their riding experience in the planetarium with what the astronauts on Mars and Jupiter's moons are experiencing during their exercise sessions. What the astronauts in space are experiencing, the students on Earth can experience with them, even communicating with the astronauts via UHF satellite–encrypted digitized frequencies.

This planetarium is one way families will become more interactively interested in outer space, including the reality of colonizing Mars, moons, and Jupiter for starters. As students, parents, and communities have educational fun in the Gravity Planetarium, they'll also be contributing to space missions. The colonization sustainability advancements for Mars must include its nearby moons for platforms enabling improved stationary communications with Earth.

Universal Space Exploration (USE) is one group of professionals pursuing private investments in outer space tourism, space mining, and planetary colonization. Launching satellites for surveying asteroids and planets for valuable minerals or resources is a prerequisite for a space mission's success. We're witnessing Earth's resources being depleted by billions of people while scientists are warning humanity of the consequences of not caring about our environment.

People willing to invest in Mars will have an opportunity to put their money in another world hosting every type of technology crucial to advancing civilization. Some of the world's wealthiest billionaires already know that space is the way to go, and they've invested billions of dollars in research and development. Like stocks, there are those who get involved early and those who are latecomers, so why aren't more people committed to being entrepreneurs in outer space?

Successful billionaires with a good education can teach, but when people don't heed their advice, they're not benefitting from all the wisdom they possess.

A hundred years ago, if someone spoke of investing in a satellite or rocket ship, they were considered crazy. Today, there are multibillion-dollar telecommunication industries and rockets going to space for numerous nations. Nations not heeding the advice of prosperous nations will be left with antiquated technologies that produce less revenue as elite nations move forward.

Global citizenship is the new avenue for progress. Space banking and space exploration are interconnected for mission objectives. Humans going to Mars, Jupiter's moons, and beyond is necessary. We must act now before humanity's resource depletion, pollution, and nuclear, biological, and chemical warfare destroy our planet. All ideas start in infancy, but with hard work, determination, and goal setting nothing is impossible.

Who can stop us if we all unite on the same goal? Space companies will work together peacefully to help humanity get to the next level. I'd like space companies to be affiliated with Colombia, South America, because of the advantage of launching into a geosynchronous orbit from an equatorial site. I've been planning for a Universal Space Exploration's mountain plateau astronomy observatory that will host a science center, airstrip, biosphere, and rocket launch site. Many investors have avoided Colombia because of the criminals, but all things will change in the near future as security improves to protect investments.

Other countries like French Guiana have launched satellites from South America. Chile has had astronomy observatories for some time.

Colombians and their children are interested in space. There will be human experimentation in close-quarter living at the Colombia launch complex to simulate tunneling conditions on Mars. Universal Space Exploration will begin with launching the smallest of rockets into outer space, engineering their own spacecraft to achieve zero rocket losses as their benchmark goal.

Peru and Ecuador may join Colombia's space program as advances are made. Schools throughout South America can send buses full of children to the space launch complex to learn about the colonization of Mars. Many nations have space programs; now it's Colombia's time to play catchup by starting its journey into the space race and by exploring our universe.

GLOBAL WARMING IS CONTRIBUTING to drought, so many territories are in severe danger of catching fire. I've personally witnessed animals suffering from the sun's intense heat. Many rural farms and ranches that have livestock, horses, zoo animals, and personal pets do not have access to permanent shade, fresh water, or nutritious land.

The earth is parched in many regions across the United States, and animals have needs that are being ignored. Forests that have been burned over the past century have eliminated so much habitat for so many creatures. We need more members of state government and members of Congress to become conservationists, funding reforestation preserves and grasslands for wild bison. Replacing the two wild bison species indigenous to the United States, which once numbered over sixty million, is a symbol of this nation's heritage.

Vast areas of America, when properly managed and protected, would be able to provide much-needed habitat for so many living creatures. For every acre of land that is burned, the federal government needs to set aside eleven acres of land for immediate protection and reforestation. It takes a very long time for a tree to grow and establish itself.

Millions of acres of land destroyed by fire still need to be replenished to provide sanctuary for numerous animal species. When animals are provided barren land with very little grass or shelter for their offspring, this puts many species in danger. Animals have to contend with both humans and other animal species predatorily hunting them and their offspring. Gone are the forests that provided safety for animals to hide among the trees, nestling down with their offspring.

To combat global warming, we need more forests, not just for ourselves but also for wildlife to prosper in a safe environment not stalked or hunted by humans. Many herds and forging groups of animals are decimated because their water has been stolen by human beings creating dams or reservoirs. Vast areas which that once flourished are now dead wastelands of burned-up vegetation. We hope that Congress will recognize the need to start a movement and a bill to replenish life in all areas destroyed by wildfires or human activity.

Local and state governments mismanaging water and taking away rivers that disperse water to regions that need it have killed much of the desert. Water that evaporates in the desert is redistributed to the surrounding plants via condensation, providing water droplets in the evening. Desert creatures that would then consume water droplets on plants to quench their thirst are left to die with no water available to them. The bees, butterflies, and various pollinating insect species of the desert, which also need water to sustain life, have had their water resources stripped away from them by human beings. With the desert parched, the cactuses, grasses, and various flowering species have become kindling, contributing to the millions of acres decimated in man-made wildfires.

The zoological kingdom is being destroyed by human beings. Native American tribes, which once had abundant water resources in the past, now have depleted water resources more and more over the years. The denial of water to Native American tribes has been an attempt to destroy their desert kingdom through a hateful revenge. Striving to eliminate or kill Native American empires through natural resource depletion is cruel, especially when nature's creatures have been destroyed on account of human greed.

When the American government cannot be trusted and its members are consumed with self-interested campaigns for their own enrichment, all citizens need to be worried about their future. People should not be allowed to own animals if they cannot provide three basic needs such as adequate access to foraging food, unlimited water for hydration, and protection from the sun. Animal shelters should also provide protection from strong freezing winds and enable animals to herd together for warmth in severe weather with their young offspring.

Of course, a new Congress bill should support all animals needing a veterinarian's vaccinations, antibiotics, and medical care along with grooming care requirements from human beings. Enforcing laws is difficult, and often people do not comply. Our bill will focus on mandatory animal care with neighbors helping each other nationwide to provide the three basic needs.

It should be a congressional bill requiring that all animals be provided a shaded canopy or awning with access to clean water and nutritious food. Human beings are suffering from global warming; we need to start thinking about animals that are also suffering. It's not just polar bears losing their habitats—the entire zoological kingdom worldwide is being harmed by human beings creating the effects of global warming.

The truth is that animals are being burned under the scorching sun, which is drying up our Earth's consumable nutrients. The grasslands in many regions are no longer green, and animals cannot consume dry vegetation in an arid environment without water. With a lack of water on some ranches and farms in arid rural regions, animals are suffering greatly, having a difficultly chewing dry cud that's not moist or green vegetation. Ask a human being to chew dry cardboard all day long without water. This is what humans are doing to animals on many American farms in the Southwestern states. These Southwestern states are desert with very little precipitation and grueling heat that can fry an egg on a rock or even the flesh of livestock. Many animals I've seen in the Southwestern United States struggle to find shade and often cannot find it while they burn in the heat of the sun. Ranchers must stop putting their animals under a magnifying glass of the sun's burning heat; the animals are being tortured.

Ask a human being to endure being in the sunlight all day long without wearing any ultraviolet sunscreen protection. Humans refuse to go into the sun all day long. So why are livestock owners subjecting their herds, horses, and poultry to the grueling heat of the sun and torturing them? Animals can feel the pain of being burned. American pets on farms and agricultural livestock are being deprived of the water they need, so they're in constant dehydration in hundred-degree temperatures.

Why are ranchers, farmers, and homeowners being so cruel to animals in their care? If you cannot properly take care of an animal, give it away to someone who can provide the proper nutrition, shade, shelter, vaccinations, clean water, and veterinary care. Also, imprisoning animals in a confined space without any other animals of their own kind to keep them company is torturous. I've seen many horses and cattle isolated in harsh conditions, suffering in the grueling heat. Are Americans crazy to be so cruel as to treat animals without any form of compassion, not understanding that animals are happy to be in social environments? All living creatures love their offspring and enjoy the company of their own kind in herds or packs. These people who want animals in the Southwestern United States need to provide more resources and mates for their animals, or they shouldn't be allowed to have them.

It's not easy eliminating the carbon emissions from our daily lives, so the creation of more hybrid vehicles is necessary. Diesel semitrucks have hundred-gallon tanks that move large quantities of freight, which keep our economy moving forward. Internationally, we need new diesel hybrid semitrucks, thereby reducing carbon emissions globally. Scientists can provide accurate data and calculations showing how the commercial trucking industry's implementation of hybrid technologies can contribute enormously to environmental improvements. Eliminating forty-seven billion gallons of diesel carbon gases from America's consumption annually is a start to reducing global warming, but only if it's implemented.

If America isn't the first nation making the initiative, then it's up to other nations to lead the way. Let's make reducing global warming a competition for young people, with prizes offered by governments. Let's start a campaign to improve all things, for we know we can do better when we communicate more efficiently.

Astronomical Investments Mission Securities (AIMS) is a new banking system for developing outer space projects. Universal Space Exploration (USE) is a unknown space group aiming for a geosynchronous orbit by launching from South America, which reduces flight expenses by half. Rockets shall be internally designed with shuttle-like qualities and variable geometry capabilities. I possess a design on just how to cut space launches by half the cost making space travel more economical. Investments in the AIMS bank will create new off-world data storage systems hackers cannot touch. Storing valuable data owned by investors within Mars in a subterranean vault will extremely limit those capable of attempting to breach private sector systems. Building the first bank on Mars to keep others' valuables safe will be part of our mission immediately from the beginning. Astronomical Investments Mission Securities is what banking should be as we elevate ourselves off world and beyond our horizon.

5

SOLVING GLOBAL HOMELESSNESS

Jesus Christ instructed all of mankind to love our neighbors as ourselves. A carpenter from Nazareth told the most elite and prestigious holy men of those times what they needed to do for the poor. Now we live in a modern age, where many elite and prestigious holy men need to be taught again what they must do to inherit the kingdom of heaven.

Sadly, in every country and in numerous island territories, there are families living without homes. Is the wealth you've acquired attributed to disobeying God? You're called to be holy, turning away from your sinful ways and striving to be more fruitful internally as well as externally toward the heavenly path of righteousness. Being rich and falling ill in the hospital isn't happiness or wealth. Not helping parents support their children's dreams but forcing them into poverty is not wealth.

What's our eternal legacy going to be? Will our legacy be one in which we neglect others, or will our legacy be one in which we bless others? What do we seek to be known for throughout the universe that is observing us and recording our history? The eyes of the Creator of the universe are everywhere, observing what we do. What do we aspire to be saints, angels, prophets, priests, deacons, bishops, popes, nuns, Vatican knights, or Opus Dei holy men?

Do we aspire to be inventors, doctors, educators, and famous actors or actresses? We should strive to keep all the Holy Commandments given to Moses. Is the title we call ourselves accurate, or are we the type of people associated with liars and deceivers, making our title less authentic? My point is, on average, we have a hundred years to do our work, which is witnessed by all the universe, and this is our legacy.

So my point is, let's make our legacy a fruitful one by first loving our Creator and all humanity. We can begin to love our Creator just by reading our Holy Bibles and filling our minds with fruitful thoughts instead of corrupt thoughts. Riches that aren't used to love or help others will eventually vanish.

I have two designs for an inexpensive housing structure to help people during normal times and now during this pandemic. I can't make it more simple than the need for equipment, land, building materials, and funding. Calculating the cost of aiding humanity by providing them shelter to escape bad weather and homelessness is invaluable. I was taught carpentry skills by my father, and though I desire to build cities or skyscrapers with my talent, my humility teaches me to build many small structures.

For thousands of years, many people have lived in tents. Also, in the Holy Bible, many different tribes lived in tents. Now, across America, many cities have homeless encampments with people living in tents. Some people in our communities are displaced by fires, others are displaced by storms, and still others are displaced by pandemics. There are people in many countries who are displaced by wars or a lack of resources. Millions of refugees live in tent encampments, and the number is rising, even in prosperous nations.

So my point is that I have many talents, and one of them is an ability to build. So I'd like to find someone who has a passionate desire to help others who can fund me doing some building to make structures for tents. Tent cities don't need to be disorganized and dirty encampments. Tent cities can be organized and well-kept enclaves in or around cities where the inhabitants have the basic necessities. Let's all do our part to improve society through peace and clean up our cities, moving homeless people to well-organized tent communities.

There will always be those who oppose local, city, state, and federal laws, but there's a reason for all initiatives. We need to help make the homeless a better environment without their destroying our environment. The airborne pollution, graffiti, waste, litter, spread of illnesses, illegal drug activity, and poor sanitary conditions in homeless encampments need to stop. Enforcing law and order is for the good of all cities and communities with families or businesses needing to recover. Homeless people are part of a temporary problem, but with guidance, we can correct them like a teacher disciplines students.

Once we get the homeless people into the communities I desire to build, then they will appreciate that the move to another area wasn't all that bad. Homeless people with the education of children can be taught small things, even after years of systematic dependency. To maintain peace, we sometimes have to remove those who aren't peaceful. Not all protestors are peaceful; some are arsonists, thieves, and violent assaulters harming others.

Since we cannot imprison everyone, we need a strategy to bring peace to bad situations. We need to protect pregnant mothers, infants, children, teenagers, fathers, and seniors in our society, especially during a pandemic. I want to aid homeless people and to ease their suffering while giving those who are mentally ill some dignity.

Our local, city, state, and federal government leaders haven't provided any dignity to homeless people in many communities. We need to treat homeless people as human beings and to establish a level of care that is fair to them. Like crying children, our homeless need a parental figure watching over them while providing basic needs. Some homeless people are groups of mothers just trying to survive and care for their children's needs.

We cannot go into church and think about going to heaven when we are letting mentally ill homeless people freeze to death on our streets. We cannot feel good inside when we have a warm home to sleep in and others around us are in cardboard boxes. Where is our compassion and love when we sit in pews thinking highly of ourselves while our actions neglect the needs of others? There shouldn't be any homelessness or hunger in this world.

Many who are rich have given their hearts to evil paths, refusing to obey Jesus Christ, but there are others who are very philanthropic. We need to get the false advisors away from our leaders and open up communication with our leaders who care. In the Holy Bible, Jesus asks the rich man to give everything away and to come follow him.

Today, there are over 2,604 billionaires worldwide and many more multimillionaires. All it takes is one wealthy person among them to fund my building these community villages for the homeless. This book makes the concern known, and it's very easy to communicate with me to get these projects started.

The love of riches is a snare that will prevent many people from rising into heaven. Jesus Christ ascended to heaven, and we're all called to follow his life teachings to be holy. To become holy is to follow Jesus Christ's fruitful paths of righteousness and visit the Holy Land for a spiritual awakening by obeying your Holy Bible's Scripture.

Many billionaires and multimillionaires will put all their trust in others until they succumb to illness, when all they had to do was put their trust in Jesus Christ for instantaneous healing. Praise the Lord Jesus Christ, for if you have faith, you will be whole. Jesus Christ told his disciples they'd be given power to heal all manner of illnesses.

Many people who love wealth will hold on to riches to the point their hearts become full of illness, consuming them unto death. Wealthy people with life-threatening illnesses who obey Jesus Christ could be blessed by a saint with the power to cure them so they live longer. Jesus Christ raised Lazarus from the dead, and saints can also raise believers from death. Christians and Catholics should choose their friends wisely.

We live in an age of many unbelievers who are faithless employees of local, state, and federal government systems. Many people go to churches or temples in these modern times. In biblical times, there were many hidden churches and temples where attendees also came together to mock the Lord Jesus, chanting to crucify him. Don't be surprised to discover that throughout the United States, there are many people employed in government who are faithless. Many of these American employees in government and the military are physical bodies and spiritual temples for many demons to dwell.

Yes, we can go to church with people, but these very people can also betray us just like Judas Iscariot did to Jesus Christ. My own family members have betrayed me, so my advice is don't let it happen to you. I believe the homeless people in America feel betrayed by their government leaders while they live in poverty. Remember that in the Holy Bible, there's the story of Lazarus, the beggar. In that parable, the dogs came to lick Lazarus's sores while he wished only for the scraps that fell from the rich man's table. In this story, Lazarus went to heaven, and the rich man went to Hades, where he suffered in agony. We have many people working in government who are rich with homes, cars, and many luxuries while there are many poor beggars in American communities.

Many men have what they call a man cave, which means the garage, a storage building, or workshop room. My project is called the CAVES RV Park, which stands for Camping Accessible Village Eco-friendly Sanctuary Recreational Vehicle Park or Community Accessible Village Eco-friendly Sanctuary Recreational Vehicle Park. There are different zoning laws for each territory regarding such parks. Camping is a more temporary accommodation, whereas the community accessible village is more of a long-term extended-stay design.

In some countries, caves can be made into camping sites or even extended-stay homes with different rooms. In the Holy Bible, when Joseph and the Virgin Mary were looking for a place to stay, they were given a cave where animals stayed. How do we solve the internal infrastructure problems in every country on Earth as we will eventually have nine billion people? Many nations are already dealing with refugees and homelessness in a disorganized fashion. The problem of homelessness has already been answered, if you look back in history. Tribes around the world have been living in tents, yurts, tepees, caves, and caravans for thousands of years.

Modernizing the living style from thousands of years ago into a new habitation isn't impossible. What started as an idea for me has become a realistic approach to helping solve the problem of homelessness. People park their vehicle under their tent's carport, and they stay above their vehicle on a roofed platform that also has fold-down storm shutters.

Living in these villages is cheap, which is perfect in struggling economies when people don't want to be living under bridges in homeless encampments. CAVES RV Parks are a stepping-stone to eventually getting back to apartments, condos, and other long-term housing for many people. Extended-stay camping is safer than living in a car. Recreational vehicle parking helps families that have upgraded to motor homes after being devastated by wildfires, tornadoes, hurricanes, earthquakes, tsunamis, and floods have destroyed their homes. If people cannot afford rent or mortgages to stay at apartments, mobile homes, or houses, they still need a place to sleep.

The tent deck village CAVES RV Park is a long-term campground upgraded to the next level above flat-land camping. Residents in my villages may restart their living in large tents fastened on top of decks with a roof over their heads. When my residents' financial situations improve, they may transition to a recreational vehicle or have both spaces if their families need more space.

If residents are comfortable in the tent deck platforms, they may opt to build a tiny home around their tent on the platform. This way, my villages provide three options for residents, suited to their own unique financial situations. Also, these villages can be mass produced to provide housing for millions of people seeking an upgrade to live off the ground and have a parking space should they acquire an automobile.

My hope is that once one village is built, then the popularity of that village will cause more villages to pop up on all the continents, helping the problem of homelessness. I believe even college students would be willing to stay at the tent deck villages while they further their education, thus improving their lives.

In village communities, there can be sectioned-off areas for singles, students, and young females seeking safety. Other areas of villages will be only for mothers with infants or young children, and families. In other sections of the villages there will be accommodations for the elderly, displaced refugees, natural disaster victims, and domestic violence support for victims. Also, there shall be sections in the villages for the homeless, recovering addicts, alcoholics, and the mentally ill or those needing transitional housing such as parolees or adult men.

The idea is CAVES RV Parks can accommodate people with many different needs physically and socially throughout different parts of the park. Each section of the CAVES RV Park will have its own safety monitoring system for residents. When many people live in a designated area, there needs to be a camera system for children playing to prevent criminals from coming around.

As with any community, no design is perfect, and there will be problems, but the problems will be less severe than what's currently happening with homelessness internationally. Local governments have failed taxpayers by not relocating the homeless to designated areas to clean up American streets. Truckers nationwide can easily deliver carpentry supplies, industrial equipment, and construction supplies to property sites for the development of these villages.

Weather-resistant decking will be built above concrete posts four to six feet deep in the ground to create the footings necessary for stable deck foundations. Tent deck villages can be built for a fraction of the cost of contending with homeless encampments which create many problems within cities. Recycled metal shall be used beneath the deck framing to provide a stronger support for the structure's weight.

Taxation without representation is a major failure in local and state governments, allowing homeless encampments anywhere they're desired. Homeless people aren't paying taxes but are living on city streets, sidewalks, underpasses, hillsides, aqueducts, and the backyard fence lines of taxpayers. Taxpayers aren't being represented by the leaders in their communities, who ignore the fact that they want safe, clean neighborhoods for their families. Taxpayers desire a cleaner environment for their families and not a backyard covered in litter with mosquitos breeding in water containers. Homeless people are contaminating environmental ecosystems littering in streams and waterways.

Why are taxpayers giving their hard-earned dollars to their local and state governments, which expose them to the dangers that homeless people can pose to residents? Homeless people have been known to start cooking fires, which have caused hillsides to burn and set neighborhoods on fire. Homeless people have been known to defecate on city streets, in business entrances, and on sidewalks used by taxpayers. Ill homeless people with dirty hands have often touched surfaces that others within the community have touched, thus spreading communicable diseases. Homeless people high on various illegal drugs have been known to assault others, commit vandalism, and even molest vulnerable children or teenagers. Children need to be protected from seeing homeless people smoking illegal drugs, sniffing aerosols, and shooting up heroin with dirty needles.

Businesses that pay local, state, and federal taxes to operate within their communities are now being exposed to infectious diseases from homeless people. These businesses, which are struggling to stay open, have to contend with homeless people who are endangering customers who have health concerns. Justice is enacting laws that help businesses and homeowners paying taxes in communities to receive preferential treatment so that homeless people are relocated elsewhere.

Since police departments, mayors, and governors in many states are failing to represent their communities' interests, then taxes should be reduced. Businesses will need to recover after the pandemic has subsided. The removal of homeless encampments within all city limits is necessary to help aid communities rebuilding and improving their infrastructures after this pandemic. Homeless people aren't paying taxes, so they shouldn't have equal entitlement to city infrastructures requiring maintenance and new development that is paid for by businesses serving customers. Underpasses that were once clean of debris or streets that were once spotless are now covered in shanty villages of trash mingled with fecal matter, fleas, rats, and urine.

Fecal waste from homeless encampments has entered city water tables, contaminating them with communicable diseases because there are no restroom facilities in these areas. Taxpayers deserve to have clean water that does not endanger their families with diseases that can be harmful. Taxpayers don't want dirty needles or homeless drug users in their neighborhoods close to their children. Dirty needles and unused traces of prescription or illegal drugs have entered water tables.

Homeless people using drugs and alcohol bring drug dealers closer to neighborhoods where they're unwanted by taxpayers. Thefts in communities increase when there are more homeless encampments because the use of illegal drugs. Under my proposal, we will aid the homeless by placing them in designated areas with housing and structures with recycling facilities.

Tent deck villages such as a CAVES RV Park can be built in underdeveloped areas where land is relatively inexpensive. Metal securement rings will be bolted onto the decking surface and wooden rails shall surround the tent fastened to the deck. Decks can be built in a variety of different geometrical shapes throughout the villages housing the homeless, poor, or destitute. When transitional housing isn't available, the CAVES RV Park will have vacancy, being the lowest tier of safe housing above sleeping on the streets.

Hip or gable roofs will be placed over the wooden railing to provide a barrier from strong winds, rain, or hail. Beneath each roof, one-inch-thick plywood panels will fold down, locking in place like storm shields to protect the tents secured to the decking. These plywood storm shields swing down, locking in place by sliding strong bolts into secure metal slots.

Tent deck villages will be best in the Southern states where winters are milder in temperature and shorter in length. Tent deck villages can be built in Northern states with the added use of polar tents and insulated panels to help retain heat generated by space heaters.

The tent deck platform can be transformed into a small home by residents purchasing the site. United Nations humanitarian programs can team up with polar tent manufacturers, carpenters, and lumberyards to rapidly build these CAVES RV Parks for governments. CAVES RV Parks can accommodate refugees, migrants, natural disaster victims, and many homeless people. The Federal Bureau of Land Management can set aside land parcels in conjunction with the US Forestry Service for emergency shelter sites for CAVES RV Parks.

CAVES RV Parks will be located in areas with minimal wildfire risk and areas needing to be redeveloped into new-growing forests. The CAVES RV Parks enable the creation of jobs for resident villagers in cleanup efforts as well as planting new vegetation. Homeless people can be trained to perform debris removal, raking, planting, tree trimming, irrigation, and other works beneficial to the US Forestry Service or fire departments.

Electrical outlets can be securely mounted to the deck post platforms to provide electricity for basic uses such as cell phones, space heaters, microwaves, mini refrigerators, and televisions. Our objective should be to treat Americans who are homeless better by establishing a standard of care other governments worldwide can replicate. In years to come, there shouldn't be any homelessness in America or homeless people sleeping on the streets. Homelessness adds to depression, which becomes one catalyst fueling peoples' use of drugs and alcohol.

Homeless people in America need to be weaned from dependency. As drug and alcohol dependency is reduced, homeless people can take on active roles in their community. Homeless people need to begin contributing to society in a positive way. Job instruction for homeless people must be somewhere on our agenda to improve communities. One of our first goals must be establishing a safe community for mentally ill and emotionally disturbed people who are homeless.

Many homeless people have been abused and taken advantage of. Many homeless people have been made to feel less human, believing they're forgotten. Sometimes the best way to correct a problem is to confront the problem. Prison is good because often it's necessary to get drug users off the drugs that control their lives and get them away from the drug dealers who control their addictions.

Prison is good if it takes a drug user out of their comfortable environment, and separates them from the like-minded addicts in their communities. America has gotten soft and doesn't lock up enough people using drugs. America lets many drug users out too soon before they make the conscious decision to change. Too many bad people are out on the streets, and they're creating chaos for decent people just trying to live their lives.

Let's feed homeless people in our communities very well in their transition to the new villages beginning their clinical treatment. A pandemic gives physicians the authority to make decisions on how to treat patients without their consent, especially if they're spreading diseases. Physicians, psychologists, and psychiatrists can treat mentally ill people best when observing them regularly in a controlled environment. Dangerous city streets are the worst conditions to begin treatment for mentally ill patients living in homeless encampments.

For physicians, psychologists, and psychiatrists to make a difference helping others, we first need to provide a safe haven for treatment. To provide an accurate clinical assessment of mentally ill patients with thought errors, behavioral disorders, and dependency problems, professionals need a monitored evaluation system in a controlled setting.

Brief interactions with patients will not suffice. There's a need for professionals to meet more frequently with mentally ill homeless people. Medical personnel need to review obsessive-compulsive personality disorder, phobias, paranoia, schizophrenia, bipolar disorder, and even psychotic episodes before a patient becomes violent.

CAVES RV Parks will help facilitate the proper diagnosis of patients after periodic evaluations from health-care professionals. City streets are the worst environment for a psychiatrist to diagnose any condition. Caring for others is the first phase of the constructive process of being a social worker, psychologist, or physician. Helping the needy to a better life needs to be a priority in America. Churches and nongovernmental agencies are helpers in the process of relocating homeless people to these villages.

Yes, there will be homeless people who refuse and may even put up a fight, but that's why physicians have the authority to medicate those who are a danger to others or themselves. Dangerous homeless people should be medicated with tranquilizers; it's a preventive measure because many illegal drugs, including meth, can induce psychotic behavior in users. Yes, there will be some people in the community who disagree with our treatment of homeless people, but they're not the physicians overseeing the transitional process.

Also, the public doesn't need to know every detail of an operation because when data is leaked it undermines a positive movement to do good for our communities. Our objective is to make America a new country, and that requires hard decisions. There are too many homeless people with fried brain cells due to alcoholism, illegal drug use, huffing chemicals, drinking sanitizers, and having preexisting mental illness.

Mental illness among homeless people just scratches the surface because there are many people who aren't homeless who also have mental illnesses. Communities across America have so many mentally ill people that some would even say American politics has some figures needing a through mental health evaluation. There are members in the US military and the federal government committing crimes who have mental illness. These criminal veteran groups lie compulsively and purposely violate America's laws and oaths.

There are those in the US military, law enforcement, and federal government who have used illegal drugs secretly but still remain in positions of power. I don't believe the public is so naive as to dismiss facts when they see poor decisions being made by representatives not benefiting the nation's best interests.

We must strive to improve America's health-care system by addressing mental illness. We need to find ways to resolve problems that have been overlooked for a very long time. If we do nothing to help others, the problems still exist. If we do something, we're being active members of our communities, participating in a positive way. Every city across America and around the world must begin corrective measures to improve conditions for homeless people. CAVES RV Parks can establish quarter lots, half lots, or full lots for each homeless person. Take into consideration that the world's population is growing and that housing everyone during housing shortages is a problem that needs addressing. There must be an alternative for those living in poverty. We must begin addressing global concerns now and not when it's too late. Empowering the homeless to feel better about themselves is the right effort. As we monitor the homeless, we can quickly find who is dependent on drugs or alcohol so we can move these individuals to another section of the community. Sections of the tent deck community can be under the direct supervision of many counselors, social workers, psychologists, psychiatrists, and religious leaders of varying faiths working together.

The homeless will be unaware they're being assessed in these tent deck communities, designed to be open in a hippie style. If the homeless feel uncomfortable, it will be difficult to keep them in these tent deck villages. We will make every effort to make the homeless feel comfortable so that they'll stay in the villages. These tent deck villages will provide great investigational grounds for the police to track down drug dealers influencing homeless people. We all desire safer communities without illegal drugs, which have given a position for gangs and violent organized crime groups, to rise up causing American murders.

Tent deck village communities can be sectioned off with particular areas assigned to just single mothers with children and small families. Other sections in tent deck villages can be sectioned off for single men, single woman, the mentally ill, and those in dependency. Residents of tent deck villages will be oblivious of how the villages are designed for certain groups. There can be no obvious distinction between the villages' areas except for the homeless to know different genders or children mean a different area of the camp. Professionals will know the distinctions for each camp section based on the individuals' personal needs.

Initial intake tents on the grounds will be a transitional area for villagers to get to know the newbies. As the newbie homeless person makes improvements, they can be moved through a upgrade by professionals into other camp areas according to their social progress. Tent deck villages can be built for prisons and immigration camps as well for housing prisoners or refugees.

Why not try to improve sanitation, health care, education, and our environment through recycling programs as part of our initiative in eliminating homelessness all by putting homeless groups to work? All humanity is interconnected with our environment, which must be protected. Getting homeless people to become more conscientiously green friendly should be an objective for all community leaders. Tent deck villages can eventually establish greenhouses and agricultural programs to help feed residents. Teaching valuable skills to homeless people is beneficial to all local communities, providing food to the needy.

To prevent abuses, a hidden camera security system can be accessed by police in these tent deck villages to help protect homeless residents from one another. Teaching homeless people skills for recycling facilities aids communities. If a homeless person commits a crime, they can continue to labor in recycling in prison. We correct those whom we care about by providing tasks for them to labor at, teaching skills that benefit communities. Modern society has automated assembly lines for cars; we should have automated assembly lines for home kits, thereby ending homelessness.

Homeless people will work once they're established in homes. If we give a homeless person a job, and, at the end of the day, they then go and sleep on the streets, they'll arrive at work the next day ill. We must provide housing for homeless people to prevent our employees from becoming ill. Prisons must become advanced recycling centers, benefiting cities and reducing pollution. Prisoners can give back to their communities in a green-friendly way sorting trash and waste products to reduce landfill excess. While earning a stipend, homeless people can labor at waste management and recycling jobs, aiding humanity and our environment.

Correctional institutions can build tent deck villages for prisoners to work at separating waste for recycling. Rail lines from cities can direct massive volumes of waste to these prison compounds for separation, classification, and dismantling. Prisoners in lower-security prisons can dismantle automobiles, electronics, and various products for recycling. If a prisoner wants to earn higher wages, they can separate valuable metals and be paid by the pound. This type of labor will be beneficial for companies needing raw metals while reducing landfills. As in freighter ports, cargo containers on the rails can be x-rayed or thermal imaged to prevent escapes from prisoner complexes.

Multiple states can funnel waste on a massive scale into a huge recycling center. At the core of the recycling complex can be a massive prison, providing the labor for all incoming waste. The prisoner complex should be in a rural area, surrounded by mountains on all sides. Prisoners will have an extremely difficult time escaping from this complex, surrounded by mountains and guard towers. As railroad cars loaded with new waste are imported into the complex, all processed waste is exported out in a continual exchange process. This is an efficient industrial recycling process for waste management that can be implemented in many territories to protect our environment. Society should no longer witness streams, rivers, lakes, oceans, and seas polluted with garbage once these recycling facilities go into operation.

Ceilings supporting large-volume tanks of compressed air funnel filtered air down long, swiveling hoses to breathing masks worn by workers performing recycling duties. Each prisoner or employee would be issued their own sanitary breathing mask to be kept clean. Employees would stand at a circular conveyor system separating waste into downward chutes. At the base of the recycling chute, hydraulic or manual pulley mechanisms would hoist steel weights to drop blocks that would crush incoming materials into smaller volumes for shipping.

The use of pulleys increases the mechanical advantage of hoisting the heavy steel used for crushing raw materials to be dropped into the chutes and weighed. Once crushed, the raw materials would be poured onto a scale recording the weight before tilting the materials into the cargo container. Eventually, after a long day of labor, each shipping container would be filled with raw materials. The recycling facilities are designed to be completely green friendly. The recycling conveyors can get their mechanical energy from waterwheels, solar power, or hydroelectric power.

Prisoners will be paid according to the volume of materials separated and weighed. Prisoners are only paid for what their own labor accomplishes in their designated areas along the separation, dismantling, and assembly line. Homeless people can be employed to picking up trash around cities, to be exported to the prison facilities for separation and processing. Prisoners and homeless people will be employed to keep America, Mexico, and the world clean.

Nations that want to get rid of their waste can pay the United States and Mexico to take their waste or process their waste for raw materials.

The most massive recycling complex in the world can be built along the border between Mexico and the United States in a remote mountainous area. The United States and Mexico can both have railroad lines transporting waste to the recycling facility, housing prisoners at various security levels.

Prisoners in super-max recycling sections of the complex can be separated from lower-security-level prisoners laboring in the same massive facility. The work areas are designed be ergonomic from the beginning entry point of unloading waste into the facility to the final exiting point of the rail lines leaving the complex. Outside the complex, there can be also another facility that can load semitrucks with raw materials for transport to other areas of the country. The prisoner complex has only one entry and exit point—a very secure tunnel—so there aren't any escapes.

The drop-off point for rail cars and trucks would be atop a mountainous plateau on an upper hillside that starts the beginning of the conveyor line. Once the materials have reached the base of the mountain, they're separated into raw form and leave via rail cars. Once they leave the tunnel, the rail cars can tilt out the contents into a loading area outside the mountain range. Gravity aids the movement of waste materials throughout all the processing of recyclables. The recycling starts high on the mountain plateau, eventually arriving in mining tunnels hosting conveyor platforms that funnel the separated waste down through the various chute systems to the base of the mountain. There, cargo containers and rail cars are filled with the raw recyclable materials.

The process utilizes geometric planes and angles to move recyclable materials through the facility by taking advantage of gravity's force acting to harness kinetic energy (Fg = mg) on raw recyclable materials. The process pulls various weights at 9.8 meters per second squared through a funneling process down sloped surfaces into separate receptacles. Every country of the world has trillions of pounds of trash and waste needing recycling. Recycling helps combat the wastefulness of humanity while providing industries with raw materials. The responsibility of all of us is to protect aquifers, water tables, streams, rivers, lakes, seas, and oceans from contamination.

Equipment used for diving operations can be reconfigured and retrofitted to provide a breathing apparatus for recycling workers. Working in an environment with unpleasant fumes requires the use of masks or helmets in addition to personal protection equipment. Protecting the environment is a global concern for all humanity.

Administrative offices have hoarded billions of dollars' worth of resources under their control, and many government programs aren't strengthening our communities. I can visit many cities across America right now and see massive homeless problems. The Department of Housing and Urban Development is a fancy cabinet name, leaving Americans living on the streets, preventing people from getting a home. Few programs exist that immediately address housing problems with implemented results.

Many homeless people have died on American streets while university graduates working in our government sleep in their warm beds with billions of dollars of properties uninhabited because of greed. I'm one of many Americans who believe the story of Lazarus from the Holy Bible, and I see the correlation in this nation of leaders who have diplomas not really making a difference except in their own portfolios. There shouldn't be any homeless people in the world, and under my administration there wouldn't be because I would pass a law to prevent homelessness.

Many countries have unemployment, homelessness, poverty, human migrations, refugees, and prisoners needing housing. At least with my tent deck villages, the poor can own something they can call their own. Government hearts do not possess a greater depth than those who are already awarded a position of authority in heaven, viewing the earth through the clouds yet humbling themselves to dwell among the unrighteous. It can only be said that governments typically do not change because of unbelief, and hearts bond to sinful ways. World history has shown that governments rise and fall. The only government that has ever lasted has been ecclesiastical polity—the ecclesiology of true saints, angelic intercessions, and faith in the Creator of heaven and Earth.

Solving homelessness also means changing the way we view homes in general. For thousands of years, homes were tents, yurts, or even caves. We can modernize these past ways of living with more modern tents, yurts, and caves to accommodate the billions of people in the world. In some countries, cargo container communities already exist.

The start-up of tent-based residential communities is a new business model that can help alleviate homelessness with low-cost alternatives by living with others in similarly difficult financial situations. Home construction programs branching into tent deck villages will be the modern model for temporary emergency housing assembled rapidly. Tent deck villages in US cities and international communities will provide low-rent accommodations to those in need. Those nations that don't live righteously will be rewarded by their own hands, so their unfruitfulness will consume them.

View America's homeless people like the baby Jesus with his mother Mary in Bethlehem, sweating, poor, and hated, needing a dry, safe place. Now imagine the rich living in affluence, thinking themselves to be royalty in the highest positions, denying any status to others. Now imagine the dirty homeless wearing their sweaty, aroma-filled clothing with stains, and a switch occurs, transferring their garments to be worn by the rich who are drunkards, blasphemers, liars, fornicators, adulterers, and sinners.

Americans in government have done to the homeless what they've also done to Jesus. The American government is living in sinful greed and hoarding riches, witnessed by the eyes of the saints, the angels, and Yahweh the Creator. The measure given will be the measure received, and to whom much is given, so much more will be required.

It's shameful that those attending universities adorn themselves with honors, high status, and superiority, but when in the government workplace, they often take credit for the work of others. The highest authority in the world has already been given to the lowliest of people living with the purest humility in monasteries, convents, and holy places, often unrecognized.

In some countries, land is awarded to each child born in their nation. This is a wonderful model of sharing what has been shared with all humanity by the Creator of this world. A nation of people having ownership in their own country is much stronger than a nation divided against itself. Citizens owning land are happier than citizens murdering one another over internal problems because they are not shared with.

Citizens murdering the police, soldiers, and veterans is a problem that will not cease and will only intensify as the unrighteousness of government toward citizens increases through persistent refusals to change.

Too many government employees, military officers, judges, and law enforcement personnel always want the highest seats of authority. Even while the world sees their many sins in the media, they still want to sit above all others. For those of us gifted by heaven who know where we stand and where we are seated, no one can uproot us because we are eternally blessed.

Our authority is from heaven above; it is not of this world. Even though others are constantly against us, trying to rob us of our heavenly rewards, we have gifts they can never wield because of their lack of righteousness. The rapid establishment of tent deck homes is valuable when thousands of homes are damaged in natural disasters.

Worldwide disasters happen in many regions. The National Emergency Strategic Task Force (NEST) can plan massive villages assembled in short time frames. These villages can last a long time, helping refugees from war-torn areas. Natural disasters need implementation plans to fix massive problems affecting many lives.

A warm, dry place to sleep is greatly appreciated by those exposed to floods, hurricanes, tsunamis, storms, and bad weather. Tent decked homes are a modern style of camping that is safer, exposing habitants to fewer insects, rodents, snakes, and water-borne organisms, which can compromise health. Providing a tent deck home to suffering homeless people touches all these areas of basic needs required in a safe shelter.

Fewer homeless people littering means cleaner oceans and communities, helping to protect the environment. A proper restroom structure near tent deck villages reduces fecal matter, urine, detergents, and chemicals entering the environment in city areas without proper sanitation. Providing people with homes reduces their exposure to diseases, germs, organisms, infestations, and weather conditions that can increase illnesses in communities when the homeless cough on city streets or touch public surfaces.

A homeless person coughing or sneezing with flu or other contagions, for example, could infect children entering school buses, subways, or buildings. Surfaces that homeless people touch, spread germs, including car door handles, businesses' doors, sinks, paper towel dispensers, countertops, and gas station dispensers.

Providing a home reduces the chance for the homeless people to expose others in their environment to airborne illnesses that affect countless cities which have employees taking sick days and reducing productivity, thus harming health systems. With fewer people becoming ill in cities across the nation, physicians will have fewer problems affecting their hospital facilities. Illnesses often lead to deaths in the elderly or very young, who are to be protected by health-care administrations.

Reducing homelessness will improve health care in cities that have been inundated with antibiotic-resistant diseases and germs, causing infections in patients that need the strongest medications. Statistical data for these cases will prove that improving one area of need, such as homelessness, improves other areas of the health-care system affected by patients' exposures to infectious, resistant microorganisms.

Not all cities are the same worldwide, with poor sanitation compromising water integrity and homelessness being major contributors to the spread of infectious diseases. A tent deck village home with mosquito netting alone can reduce many diseases spread over vast areas in poverty-stricken regions, thus saving lives. Infant mortality can be reduced through tent deck villages, as can other deaths attributed to bites or earthen exposures.

A home enables homeless people to better organize their mental state, reducing anger and thereby reducing violent crimes and thefts. A home enables homeless people to organize possessions, improving new skills and learning in a safer atmosphere than living on the streets, constantly in danger, being vulnerable to others or the environment. A home reduces mental illness and feelings of being vulnerable to insects, rodents, weather, and dangerous people, thus helping the progress in reestablishing communication for improving therapy for those in need.

A home helps improves the education process to determine if someone is hoarding because they cannot read while belongings help them feel like they have more value. A home enables the homeless to sort their belongings, thereby reducing litter. Homeless villages can have a recycling program in which money is earned in exchange for work separating plastics, glass, cardboard, and other items, teaching new skill sets to homeless families.

Homeless people can learn new skills and contribute to the society of all nations that need to recycle. A home reduces litter and increases recycling in an organized fashion, benefiting all. Worldwide, eliminating homelessness must be a United Nations consolidated effort to deal with many natural disasters, refugees, wars, and misallocations of funds not reducing poverty in a sustained, substantial way.

What needs to stop is the suffering that children, women, and even men experience on cold streets, exposed to thunderstorms, floods, and extreme heat combined with humidity, which produces insect pestilences. Refugee families need new designs to temporarily lift them out of bad weather conditions in many territories. Bad weather affects everyone, including the homeless, who are often tormented by greater suffering after floods destroy the places they have resided. When the homeless are helped, sanitation conditions improve in many regions, reducing the onset of diseases.

My business model is that if mobile home parks and apartment complexes cannot accommodate the homeless, I can with the help of others who care enough to aid humanity. Governments can provide funding for my tent deck housing villages, which will provide housing for the homeless at the lowest income levels. Governments seeking an alternative to Section 8 housing or welfare programs can consider the tent deck housing projects. Where charities like the Salvation Army, Catholic Charities, and other groups are unable to accommodate the homeless, my villages would meet this need. Tent deck villages would provide an alternative to the cold streets for children, women, men, and families.

My tent deck housing villages would be able to accommodate the greatest number of homeless people, always having available vacancies. Government grants and private investors can help me establish the first villages, which I would be willing to build with my own carpentry skills. The government welfare program can pay me for the lot rent at the lowest rate in any territory in the United States or internationally. Money allocated for refugee housing can be given to me to establish these villages in the United States or internationally. Catholic Church volunteers skilled in carpentry would be able to provide the labor to erect many villages in areas of need.

Cities with minimal resources to house the homeless can look to me to meet the demand, providing tent deck housing structures on private, city-owned, state, or government land. Fields and open land spaces, once unused, can now become tent deck villages, addressing homelessness and inviting social workers and outreach programs to actively participate. The villages will be open to professionals visiting to help the homeless and to bring religious and government professionals to assess the conditions.

The villages are really a place to sleep in safety, so the homeless have minimal belongings but are not out on the streets suffering. Rules against hoarding will be established in the tent deck villages to keep the areas free of clutter to beautify landscaping, greenhouses, and village gardens. Each homeless person will get their own plot to call their own. Cameras throughout the village will view all areas, as theft will not be tolerated.

Like any program, there will be those who do not follow the rules, but the emphasis will be no one sleeping on the streets. For those who do not follow the rules there will be areas set aside for problem makers and mental health evaluations. A section of the tent deck village will be assigned to vehicle-owning residents putting their lives together. Mothers with children improving their lives may eventually move into apartment complexes or Habitat for Humanity homes once they progress in the program.

The villages are really designed to help people get back on their feet and to make sure no one is on the streets but rather in a warm, dry place. Church congregations, city offices, and government offices can bring meals to the villages, much like a food truck soup kitchen. Decks can be erected in different shapes with themes that are culturally significant to the regions where the communities are erected. We can provide housing to refugees, but we haven't provided housing for American citizens; both areas need to be corrected.

Tent deck villages in a variety of different themes will make the experience unique in each city, much like hostels. Tent deck villages erected for the purpose of providing low-cost camping similar to hostels will be another option to enhance international travel, improving global socialization. If a tent deck village's tenants can't pay because they're indigent, they can still stay at a very reduced rate, which the government can supplement to keep people off the streets. Government welfare can pay for this type of inexpensive housing for those who register for assistance, with no one being turned away. No human being should be made to sleep on the streets like the rats that frequent storm drains, and government has failed many people.

Villages will consist of nice tents on decks in rows with a very nice restroom/bathing facility shared by all residents. Governments have collected trillions of dollars from taxpayers, and social workers haven't improved conditions for countless American citizens. One way the government can correct its errors is by working with me to establish parks full of hammocks or hammock tents suspended between concrete posts to provide free sleeping areas for the homeless. Tent deck villages give a level of humanity to the homeless, allowing them to be treated with dignity and not exposed to the elements.

City shelters unable to accommodate the homeless because of a limited number of beds can now create parks to help the homeless have a place to sleep. Two vertical wooden or metal poles can be inserted into holes filled with concrete. Between these two poles will be a third pole, secured horizontally between the two vertical poles, creating a rectangular shape. Between the two secured posts, a hammock will be secured in a level position. The horizontal post can be secured in such a way as to place a small gable roof over the top of the posts, or an awning can be installed to protect the hammock from strong sunshine, snow, or rain.

The advantage of these structures is that mosquito netting can envelope the structure, protecting the person sleeping inside from pestilence. Since insects transmit diseases, these villages and parks can be beneficial to health departments and hospitals, controlling or treating disease. Rest stops could also establish these structures in remote areas where hotels, motels, or shelter isn't readily available.

Awnings over the hammock structures could have small to medium-size solar panels to generate electricity for LED lighting, in addition to having duel outdoor outlets for powering phones and smaller devices. There should be adequate electricity to operate small radios and televisions to keep inhabitants informed of emergency news, entertainment, and educational programming.

Maintaining peaceful park facilities will be easier than dealing with homeless communities scattered all over cities. Often, homeless groups will erect structures in areas where their presence is undesirable for tourism and families seeking safety for their children. I have many years of construction experience and can get these villages built. Like many projects, it comes down to resources and land to start the development.

I can recruit many volunteers to help me with these tent deck villages in ten cities to be the first communities built. If government leaders reject me, they must give account for themselves in the afterlife for opposing charitable works. I have a plan; what I need is financial support, land, and the equipment so that I can put these projects into motion. Daily, millions of dollars are spent on courts, national and international investigations, fraudulent investments, false charities, and countless other programs, but why haven't leaders aided me?

Reject the spirit of Jesus Christ's servants, and that is what you have as your testimony before the face of Yahweh the Creator as your life's legacy. The American government has been looking away from the problem of homelessness for a very long time, so it's grown exponentially. The financial distress of Americans has been ignored intentionally by the government to increase harm upon citizens rather than alleviate the burdens of citizens paying taxes. Annually, homelessness continues to increase while the empire of the federal government arms military families with the resources to oppress even its own communities full of suffering citizens.

Philanthropists are the ones who will be the most remembered in world history because they're the ones funding the projects that help alleviate humanity's suffering. Carpenters from ten Catholic churches in ten cities can improve Los Angeles, San Francisco, Austin, Houston, Portland, New York City, Seattle, Las Vegas, Honolulu, and Washington, DC. Why are Catholics not believed? Catholics have built cathedrals and skyscrapers with the professionals in our congregations, while aiding many through charities. The National Emergency Strategic Task Force (NEST) is an intricate part of our Lord Jesus Christ's crown. We are all called to do fruitful works to love others and end homelessness.

Catholics need to unite with me to accomplish ten projects to be an example that we can do what governments aren't doing with their immensely huge budgets being mismanaged. I need funding to support the project. The Holy Bible says to ask, seek, and knock—exactly what I'm doing here for the purpose of building these projects. Helping the homeless in ten American cities is a start; we can see the before and after pictures of how we made a positive change.

My father is a devoted senior of his church who taught me carpentry since my youth. The biblical story of Lazarus has inspired me not to ignore others in need. Homelessness does not need to exist with nations possessing so much wealth and powerful military governments. Governments have refused to do what they should have done already. Long ago, all governments should have addressed the many problems associated with homelessness in their nations. Many governments globally haven't been raising their people up to a certain level of dignity, with permanent shelters to dwell in.

Tent deck villages give a level of humanity to the homeless and treat them with dignity, not exposing them to weather elements or pestilence. Tent deck villages can be transitional housing for prisoners. Disadvantaged people can benefit from the CAVES RV Parks, which provide immediate housing in emergencies. Even soldiers doing humanitarian aid in foreign countries could erect tent deck villages or hammock tent structures for their own housing during their long-term assistance in catastrophes. A proper place to sleep is a crucial component of the health-care system for all citizens, as is proper nutrition.

Tent deck villages can have strict rules like no junk accumulation outside residences and no excessive hoarding in village dwellings. A housing association can help enforce rules to keep the village communities orderly and clean. Just having a place to sleep makes people happier, and to deny them this happiness is a cruelty.

The United Nations needs to stand against governments practicing cruelty toward their citizens by not providing the proper places for their people to sleep. Many years of life are spent sleeping, reducing fatigue and supporting a healthy metabolism, proper digestion, and good mental health. What do visitors to America see traveling across the country for the first time? Many people in emergencies travel with minimal resources, fleeing terrorism, crime, criminal violence, hurricanes, tornados, floods, fires, pandemics, storms, and earthquakes, so they must find safety.

6

WE MUST REVIVE OUR ENVIRONMENT

Our environment's dead trees must become part of an inexpensive funeral service for those desiring cremation. Our society, which has neglected our environment's need for the water of life, needs a revival in spirit. If we shed tears in our forests, that's the catalyst for new life. Today, forest fires often get out of control because of all the accumulated branches and brush filling spaces between acres of trees. Funeral pyres would be made primarily of fallen tree branches from forests and trees taken down in preventative maintenance to stop forest fires.

We need to stop massive fires having excessive material for consumption, which increases danger near family homes and property. The "fallen trees to living trees" concept helps alleviate excess by collecting wood for transformation into ashes to become new forests. In droughts, when branches and brush become combustible, removing these materials for funerary services helps the forestry service. Retrieved cremation ashes, dust to dust, are ceremoniously buried beneath the roots of newly planted trees and flowering seedlings.

Cremation ashes are also used for seeds sown around each tree. A plaque or marker with a name can be placed with each new tree. Families can request the type of tree for the funeral, with many species available. Churches can establish greenhouses for this purpose, and the minimal cost for the funeral service will help maintain the trees grown by professional arborists teaching youths this skill within churches.

Jesus Christ was crucified on the temple's last stock of wood, a rejected tree; we are to continue to carry our crosses even when rejected. When the Sanhedrin and Roman soldiers mocked Jesus, the rejected tree lumber was chosen for the cross to crucify our Lord Jesus Christ, ridiculing him as King of the Jews. Loved ones who have passed are remembered through new life, by the emergence of new tree growth, which can grow for thousands of years, becoming a peaceful sanctuary.

Forests already burned down and needing reforestation can serve a new purpose, becoming cemetery sanctuary forests with caretakers for those resting at peace. A new path: a celebration of life ceremony rather than a traditional funeral. Walking through trails full of beauty is a great way of remembering loved ones, witnessing the beauty in flowers, trees, and living creatures abiding with the deceased like a Noah's Ark forest. Many older cemeteries have little space to accommodate new coffins.

What's wrong with burying a deceased person's coffin in a vertical position, a seated position, or in an Indian-style crossed-leg position? The deceased loved one's spirit has arisen to a new life, and the physical body will become dust. Ashes to ashes, dust to dust, as is often recited by priests, reminds us that our physical bodies are only temporary dwellings before we rise to a new heavenly life. I actually like the thought of the heavy headstone being placed directly over a loved one's vertical grave so that the plots are square and protected. The visitors to the cemetery will not be standing above the deceased person, for the headstone helps protect the coffin beneath it.

Burying loved ones beneath the headstone also will enable a reduction in expenses, as less land will be required for the burial for families struggling with many expenses. Maximizing the use of cemeteries for hundreds of years is beneficial to communities with growing populations.

If deceased loved ones are buried in seated positions, then the coffins can be smaller, reducing the amount of materials used. Also, materials once unusable for coffins could be usable for the smaller size, thus maximizing usage of supplies. Some monasteries provide labor for making coffins, and with many forests being depleted by wildfires, stretching supplies can be good, especially during times of uncertainty. So many countries didn't expect to deal with a pandemic, and now they must find efficient ways to bury their dead loved ones in a respectful way. History tells us that France stacked their deceased within the catacombs because of limited space. As large as America is, just being more efficient in the way we bury our loved ones can be beneficial not just for ourselves but also for the environment.

Trees have been around since the beginning of the earth, so planting new seedlings like cedars, cypress, palms, pines, boxwoods, olive, redwoods, sequoias, oaks, cherry, and acacia will give families a forest to visit. Families entering the forest can peacefully reminisce about those they loved while abiding in nature. With such a broad assortment of species to choose from, including biblical trees, seniors can choose new growth according to their faith. The voice of one crying in the wilderness prepares a place for our Lord Jesus Christ.

New tree growth emerging from our ashes aids the production of oxygen, purifying the earth and beautifying thousands of acres of lost forests. Forests devastated by wildfires need caretakers, and the government isn't able to govern all those lands because they allowed the forests to burn to the ground. Prisoners need to work, and getting them planting prairie grasses, flowers, shrubs, cacti, and an assortment of trees is good labor. Churches and nonprofit organizations need to consider the adoption of burned forests. We have the adopt-a-mile program on highways; well, let's have the adopt-a-burned-forest program. The creation of reforestation programs for memory gardens serves many communities that need cemeteries different from the current design. Narrow walking trails through memory forests will encourage youths to visit their loved ones while promoting cardiovascular health in addition to combating heart disease.

Stations of the cross can line the narrow trails winding through the forest behind celebration of life structures. Lining the peaceful forested pathways are concrete benches, sitting stones, wooden benches, wildflower gardens, and rain shelters. Families can plant flowers around a loved one's tree. Hummingbird feeders can be hung from the branches of the trees. Varying geometrically designed planting patterns from one shape to another can make each cemetery different yet beautiful to visit.

Replacing fallen trees with living trees creates jobs: arborists growing trees, gatherers collecting fallen wood for funeral pyre assembly, forestry crews clearing fire hazards, sanctuary forest landscapers, and controlled funeral pyre maintenance crews. Civilizations must care about forests, and this program can be for conservation. Many of these forests will become homes to many living creatures including birds, flying squirrels, and deer.

Erected celebration of life reception centers for family services can include PowerPoint presentations, videos, photos, banquets, and assorted activities to remember loved ones. Heaven is where the treasure is; this is where God calls us to place our thoughts because we all pass away. Let us rest in peace in a beautiful forest with God's creatures and children walking through our nature trails visiting us as we rest. The funeral pyre can be floated out to the center of a lake on a barge; a fire then ignites the wooden structure, starting the new life transformation service.

Another way to use wood taken down in forests for preventative maintenance is to give the lumber to monks who can transform it into miniature caskets for fetuses. Abortion is cruel, and fetuses need to be placed in miniature caskets that can be placed in a mausoleum wall. Caring for the deceased is an important part of humanity. When humanity stops caring for the deceased, then its morality turns to darkness, for even Jesus Christ was cared for at his burial. All societies need to care more for each other, including the deceased, with genuine remorse for the loss of innocent life.

Burial methods must change to reduce the amount of grave space in cemeteries by repositioning the body in a seated, standing, or crossed-leg-style position. Why can't coffins be vertical and the head stones cover the perimeter of the casket? A square over a rectangular coffin would also help protect the coffin itself. A vertical burial is a vertical resurrection and ascension to heaven. These options for burying loved ones can also provide cemeteries a way to maximize space or reposition family plots.

All abortions need some sort of remembrance, and I believe miniaturized coffins will help remember life that Yahweh gave. I do not support abortion in any way, but we have to start the caring process for remembering all fetuses. Abortion is cruel, malicious, and heartless, doing harm to miracles, so we need miniature caskets to wake up mankind that murdering fetuses is unacceptable. Usable wood salvaged from burned forests, like sequoia and redwood, can be used to make tiny caskets for fetuses. I'd love to see wood from burned forests used as furniture in a monastery to remind priests that they must increase the life of the forests crucial to all ecosystems. Priests and monks in monasteries are peaceful, so the growing of trees is a happy activity for those who love all the miracles of life. I've spent the time to design a cemetery, and now I believe it's time to redesign the whole schematic to cater to the needs of our environment.

Ornate cemetery walls can be erected to hold each tiny casket, remembering the lives lost in over a billion abortions. Society as a whole needs to feel more remorse, care, and the need for prayer to acknowledge the problem in the hearts of many. These miniature caskets can be also inserted into concrete materials added to an ocean cemetery reef. Over time, artificial cemetery reefs will become living reefs with a variety of anemones, coral, fish, and other marine species caring for the fetuses as guardians of a peaceful place.

With humanity destroying coral reefs worldwide, it's only right to attempt to rebuild new reefs to combat the ignorance that has led to this destruction. We breathe oxygen provided to us by nature from the oceans and from the living species on the land. It's our responsibility to facilitate new life in both these areas for all future generations.

Building underwater cemetery pyramids is another concept for those desiring to be buried in oceans, lakes, or seas. These cemetery designs will help facilitate new growth by aiding in the regeneration of coral reefs. Cremation ashes are mixed with aggregates and poured into molds, hardening like concrete to form various shapes for placement on the ocean floor, creating new artificial reefs. We all need to improve our work protecting Earth's creatures threatened with loss of habitat or extinction. We must come up with new ways to deal with the internal collapse of Earth's ecosystems, working to aid nature so it can flourish again.

Federal land and grant money could be used to start cemetery forests or artificial underwater cemetery reefs. If the government doesn't want to participate in these developments, churches and private groups are welcome. The only requirement must be to acknowledge belief in the Creator of heaven and Earth to be accepted into a cemetery forest or cemetery reef. We live in a world with many religions, but these cemetery grounds will be strictly for those who believe in one biblical Mount Sinai Creator of all things in the universe.

In America, there are many states that bury unclaimed bodies from nursing homes, hospitals, and even prisons. I'm suggesting the establishment of a penny fund project in which pennies collected throughout a person's life can fund a proper burial. America has vast amounts of Bureau of Land Management property, including desert areas, which could become cemeteries. Americans born in the United States should be given a basic headstone and a burial plot, even if they cannot afford it. The Southwestern United States has mountains of rocks to be headstones.

A few bags of cement and some stencil lettering isn't much of an expense. Also, prisoners can easily dig graves and bury the deceased as part of their service to the prison system as well as to the federal government. There can be one or two sites made into national graveyards for unclaimed Americans. The Tomb of the Unknown Soldier honors veterans; well, let's give Americans who've passed away poor and forgotten a proper burial.

Many states have desert property where diseased and deceased bodies can be best kept in a remote location. During a pandemic, bodies that are diseased are best kept in one location to hinder the spread of infectious disease. Refrigerated semitrucks can deliver diseased bodies in labeled cardboard boxes to cemetery sites for burial and cremation. Billions of pennies collected can go into a treasury account for one purpose: to provide even the poorest Americans a proper burial and/or cremation.

Aiding me will also aid a monastery serving the Lord Jesus Christ with fruits of the Holy Spirit. Assembling religious leaders to feed the hungry, shelter the homeless, clothe the naked, give drink to the thirsty, heal the ill, visit the homebound, educate others in the faith of Jesus Christ, and promote world peace should be an investment we all make.

All nations need to have concern for the environment, because every country I have visited has a problem with trash, and this originates from the ignorance of human beings not caring about their surroundings. Pollution doesn't just harm humans; all the other species on Earth are being affected, and there needs to be harsher punishments for those behaving worse than animals by littering. The chemicals in plastics, paper, and other materials all cycle through our environment, returning to humans. It's not just big corporations that pollute; it's all the billions of humans not tending to their waste in the correct manner: picking it all up and disposing of it correctly.

Too often in our communities I've seen litter from humans in alleys, gutters, parking lots, roadsides, stream embankments, riverbanks, waterways, and coastlines. The wooded areas in communities have huge accumulations of trash from homeless encampments. Nationwide, all local governments need to start enforcement, however unpleasant it maybe for the homeless, to stop this horrible behavior. The homeless often have mental illness, and the best way to teach the mentally ill is repeated instruction like good parenting.

The scattered litter from communities is entering our oceans at an alarming rate. It's created major environmental problems, destroying ecosystems. Environmental pollution has a cascading effect on surrounding ecosystems, including wildlife. Wildlife is dying because human beings haven't united to clean up the environment. The root cause of wildlife endangerment originates from homeless people contaminating areas with pollution and humanity not placing litter in the proper waste management receptacles.

The whole world needs to teach our children to pick up trash and dispose of it correctly. Those who litter need to be treated as the lowest class of living beings, even below the earth's animals, who do not litter. This is the only way to address the massive environmental problem: providing a comparison to establish the elite from the corrupt members of society.

Being nice to people who don't care about the environment is the wrong approach to correct the global litter problem. Alienating the mentally ill into their own groups does not correct the global litter problem. The answer to the problem is punishing those who litter by requiring them to provide community service, cleaning up the litter until they develop calloused hands. Teach a person in error once with harsh punishment, and hopefully you will not need to repeat the punishment because they've been taught right from wrong.

Too many governments are cowardly, not enforcing necessary punishments that are crucial for taking care of Earth's creatures in their natural environment. Governments doing nothing to stop an environmental collapse have inadvertently destroyed Earth for future generations.

Creatures that thrived generations ago have become extinct because of the failures of humanity in earlier generations. This catastrophe is a direct result of human beings not caring about the needs of all the earth's creatures, which includes all forms of life. The ecological interdependency of creatures in their natural environment is very sensitive, and humanity has been insensitive. When humanity seeks cures for diseases in the future without finding them because animal species are rare or extinct, governments must accept their failures. Governments failing their citizens to the point that they become ill and perish are the nations that current generations of young people desire no leadership from.

Society needs to get the homeless working and cleaning up each community. The abundance of trash in state, public, and federal parks is a problem. Punishing the homeless is not bad; it's called correction. Homeless communities in poverty contribute to international problems because of the increased introduction of pollution into the environment. Homeless communities in poverty contribute to the spread of sexually transmitted diseases and the increase in illnesses caused by poor sanitary conditions. Governments must overcome homelessness to facilitate a healthier environment for citizens.

In past generations, young people were given challenges in their communities to improve the environment. Boy Scouts and Girl Scouts can be given challenges individually or compete with other troops to win awards for collecting the most trash by volume and weight. Possible prizes for the youths participating in the challenge: a game system, a trip for the troop somewhere, or even a cash prize.

This environmental challenge can also be presented to public and private schools to help keep America clean. Government grants can be awarded to schools educating students on environmental concerns surrounding health and protecting our national parks. As more of nature's creatures die, including amphibians, fowl, and other organisms, students need to fix these problems by getting involved in green movements.

Trees that have died from nearby pollutants have been filling our forests with kindling, which adds fuel to the forest fires threatening our communities. Poor communities with outhouses need to use dead kindling beneath excrement, which should be burned periodically in a metal bin prior to discarding. This is to prevent contaminating tree root systems and water supplies with bacteria or viruses from excrement. The ashes can only be introduced into composts after being processed by intensive heat. Students today are less concerned with their environment, so stimulating this concern early in life is crucial to protecting Earth's ecological ecosystems.

The US Fish and Wildlife Service and the Environmental Protection Agency need this generation of young citizens interested in learning about environmental protection concerns. Storms that inundate areas with landfills are problems since runoff waters contain an assortment of contaminants. Outside landfills after major destructive storms, chemical pollution from leaking containers seeps into flooded neighborhoods.

The combination of landfills, neighborhoods, sewer systems, treatment plants, and industrial facilities losing chemicals that mix after major floods entering nature is an environmental concern. Oils, bleaches, detergents, pesticides, insecticides, gasoline, fertilizers, solvents, lead paints, pharmaceuticals, poisons, and assorted substances from canisters affect our environment and endanger humans, organisms, and animals.

Exposure to airborne chemicals and soil and water contamination affects not only nature but also animals and humans. Numerous medical cases and lawsuits have proved that contaminated water causes many diseases, mental retardation, poisonings, miscarriages, birth deformities, and even deaths in humans. Currently, poisonous fumes exit the ground baking beneath landfills in strong concentrations.

The parts per million (PPM) concentration of toxins in the air is harmful to humans and animals not often issued self-contained breathing apparatuses (SCBA). Employees in many of these landfills do not have the proper personal protection equipment like rubber gloves, wader or knee-high rubber boots, air particulate breathing masks, and protective goggles. Damage to the nasal passages and respiratory system is a factor for the animals or humans near these contaminated sites without SCBAs.

Airborne asbestos particulates in landfills are dangerous to the eyes, mouths, and nasal passages of employees and nearby citizens. The epithelium of the body and the membrane of the upper respiratory tract are vulnerable to toxic gas exposures that can create infections or neurotoxicity, affecting brain activity.

Improving landfills also means reducing waste and the chemicals that seep out of them. Toxic black fluids escaping from beneath landfills can create cancerous growths; these chemicals must not enter underground water tables. Standing water accumulates at the base of these landfills. Hours of exposing employees to carbon monoxide, methane, and other toxic gases must cease.

OSHA needs stronger oversight for the well-being of employees to ensure they're not exposed to chemical gases that can induce headaches, nausea, illness, and respiratory problems. Many of these toxic gases can damage the brain, altering thoughts and or intoxicating employees with mind-altering toxicity. Fluids from landfills vaporize, and the chemical concentrations in the air also adheres to water vapor in clouds, which redistribute the toxins elsewhere.

Preventing chemicals from underneath landfills from entering the atmosphere helps improve health by preventing cancerous substances from becoming airborne and being redistributed as chemical-saturated condensed water droplets. Waterways with these chemicals entering our crops and food supplies endanger human health. More toxicological analysis by the Environmental Protection Agency (EPA) is needed to measure gases emitted from landfills at various heights and depths.

Research needs to be conducted on how these landfill vapors enter our upper atmosphere, including the clouds and wind systems above our cities. Children eating snow is not advised, in addition to not physically entering waterways after rainfall for their protection. Chemicals from the upper atmosphere can mix with rainwater and create a catalyst for bacteria to change, becoming more virulent due to the toxicity. Bacteria and viruses want to live, so they adapt to their surroundings, and if their watery environment is filled with toxic chemicals, these organisms change according to their exposure.

So when children die unexpectedly from rare bacteria in the water, how did the bacteria arrive in those locations in the first place? When bacteria evolve because of chemical changes in their environment, then the danger can be much greater. Chemically resilient bacteria can attack our aquatic food supply and land-based agriculture. It's not just the livestock, poultry, and crops that will be affected by these bacteria; it's humanity struggling to survive as a result of the neglect of the environment.

The way trash is processed needs improvement. I will patent a process in which tanks of breathable air supply self-contained breathing apparatus masks in a series from a warehouse ceiling. The recycling facility will overlook a desert valley from a mountainous incline with a series of channels that help separate recyclables that slide down corridors to the mountain's base. Inside the long warehouse on the mountain above is an area where dump trucks unload their trash.

The trash entering the complex rolls across a winding conveyor belt at which numerous prisoners in SCBA masks stand, separating trash. The prisoners are issued PPE, which is washed daily after their shift separating trash. Twenty-four hours a day, 360 to 365 days a year, minus holidays or grace periods, prisoners work in rotating shifts separating trash.

Around each prisoner is a series of chutes, each of which is designated for a different material. Prisoners are paid according to the weight and correctness of the trash separation, a stipend for institutional commissary use. The goal is 100 percent efficiency in recycling so there is zero waste between the United States and Mexico.

Worldwide, the United States and Mexico will take the position as the best countries at recycling, even cleaning the oceans of waste. America and Mexico will lead the world in reducing the environmental impact of waste by cleaning the oceans effectively and processing the waste for the international community. The system of conveyor belts moving along the winding track through the warehouse provides a center area for each prisoner to stand. Conveyors carry new waste brought into the building, but prisoners breathe fresh air in their masks. All forms of waste are separated in the new recycling system, including plastics, glass, paper, metal, liquids, and organic materials. The chutes funnel materials into a small container, which is compacted by hydraulics, and weighted mechanical pulleys then dump the contents into a larger container, once verified for consistency. There's a field of containers at the base of the mountain, where the materials go through a fenced-in three-layer railway or semi-inspection area. Infrared devices scan all material containers, making it impossible for prisoners to escape from the complex.

The ergonomics of the landfill facility from entry to exit are designed for easy-flow operation with much attention placed on design, security, and efficiency to process the most materials effectively. From the ceiling, long hoses extend off swinging arms attached to a central support cable to which the prisoners attach their helmets for breathing in their designated work areas. All prisoners go through metal detectors and inspections to prevent illegal items from entering the prison facility. This recycling facility needs people wholeheartedly immersed in designing the best structure with a passion for doing an excellent job that's worthy of the history books. The recycling facility must be built to last for hundreds of years, not quickly assembled because of bidding wars to obtain government money as an easy paycheck.

The United States and Mexico share the same problems—drugs, violent cartels, gangs, terrorism, and corruption—so let's build a megaprison for all the criminals to be shared on the border. Forget the border wall; create something useful so prisoners can labor in trash recycling, separating materials for two countries for environmental cleanups. Criminals ruin the lives of children, mothers, and families. Morgues are full of drug overdose victims stacked from the floor to ceiling while drug dealers are free, needing to be imprisoned.

Birth deformities are on the rise because of drug use. Fetus and infant deaths are also rising in addition to premature births because illegal drugs are poisoning many mothers from many nationalities. Infants born addicted to drugs with developmental problems and mental illnesses are a massive problem in the Americas. Drug cartels murdering innocent people in Mexico and the United States are a huge problem. Violent gang activities must be stopped by imprisoning these criminals who are harming our societies and our peace.

Building a megaprison to incarcerate criminals is a much better usage of billions of dollars than a border wall that Mexicans will only dig under or catapult over. The Mexicans have already proven themselves to be efficient at tunneling, digging elaborate systems to smuggle drugs into the United States. Prisoners doing the hard labor of recycling will offset the facilities' operational costs. Instead of a high wall, some areas can have surveillance towers with long-distance infrared capabilities that scan quadrants through telescopic lenses.

Officers can have constant observation from the towers while having digital marking systems in quadrants between all borderlines and spaces. Companies needing resources can buy recycled material from the megaprison for their manufacturing operations. There will no longer be waste because of these recycling facilities, and now nations can recycle everything efficiently.

Four US states border Mexico, and each one can benefit from the megaprison sending prisoners to work. Mexico also can benefit from the megaprison doing recycling. All US territories can send their state and federal prisoners to be housed in this megaprison. Organic materials can be used in greenhouses and in farming, processing materials properly after all guidelines are observed. Organic materials not used for food production may be used for reforestation projects or planting assorted species for city or highway landscaping. Rare endangered tree species can be planted and raised at this recycling facility. Growing things can be a good hobby or job for prisoners with life sentences.

The United States can establish a mutual cooperation with Mexico's military and federal police, imprisoning all dangerous criminals with life sentences. Incarcerating the dangerous criminals of the United States and Mexico is a mutual need. The super-maximum-security megaprison is something both nations need to deter those not afraid to murder the police. Both the United States and Mexico can employ labor to build the prison. Both nations will pay for the megaprison to be built so it can be enlarged in future generations.

Population growth is a worldwide factor to consider when outlining the expense for this megaprison project. Hiring an architectural firm without expertise in prison design is a bad idea. All the best ideas for the design need to be addressed, with all political parties coming to an exhibition revealing designs. The functionality and ergonomics of the prison must be addressed, and only the best materials must be used in the construction of the megaprison to enhance its longevity.

Some sections around the megaprison can be tent prisons and medium-security prisons that aid the maximum-security prison at its very remote location with one road in and one road out. The idea is for the United States and Mexico to share the cost as well as the personnel of this megaprison. The message to criminals in both countries will be that no one is putting up with their criminal behavior.

Mexico and the United States will operate the prison under joint military/federal police control, with each nation's president selecting the staff. The presidential executive office will authorize the detainment of terrorists, cartel members, organized criminal networks, gang leaders, serial killers, and all criminals sent to this prison facility as threats to national security. Both countries shall share control of the megaprison. There will be two world leaders from two internationally recognized nations shaking hands and working together to combat crime with zero tolerance. That's what America needs to show all nations that it will not tolerate criminals in the drug industry getting away with whatever they want. Many nations refuse to do the right thing out of fear, but America isn't afraid to go on the attack, willing to uproot all the criminals for imprisonment.

Both nations with improve the environment using prisoner labor. Drug dealers accustomed to the high lifestyle can now play in the trash as a portion of their punishment for harming the lives of people in society. Prisoners housed at the megaprison will perform various forms of recycling at the landfill adjacent to the prison in a heavily guarded mountain range. Observation gun towers will surround the mountainous perimeter of the prison facility, including its landfill processing areas. Guards can wear masks with air supplied to reduce the odor of any fumes from the landfill when outside their self-contained bubble towers.

The designs for the landfill operation are a hundred years more advanced than current designs. The development of the megaprison project will create good jobs for the United States and Mexico working together as an alliance against criminals. The development of this megaprison will place fear in criminal networks because there's ample space to house them all. This will drive the criminal networks into clandestine mode, and that's a perfect way to find all the rotten thugs for biometric profiling, ready to be arrested.

Police officers and soldiers injured in war can find employment working in control booths, gun towers, inspection areas, and various jobs throughout the prison. Tough, physically handicapped police officers and soldiers can continue serving inside the megaprison so they feel useful as productive members of society. Criminals will be in fear of being sent to this maximum-security facility. Criminals of every kind can be housed in the megaprison, including illegal drug manufacturers, sex traffickers, drug lords, narcotics dealers, thieves, murderers, terrorists, rapists, child molesters, cannibals, fraudsters, white-collar criminals, and members of organized crime syndicates.

I WOULD LIKE to propose that the Labor Day weekend be extended for Americans. Traveling across this nation can take many days by vehicle. For many families, one weekend does not provide much enjoyment or relaxation, especially for parents. Parents spend the weekend loading and unloading the vehicle as well as preparing meals. Half a weekend driving in state or across state lines makes the Labor Day weekend itself labor intensive.

What I'm proposing is a national American Families Traveling Week or Recreational Vehicle Week. This will be a time for families to take their children on a small vacation across this beautiful nation. This week should follow Labor Day so that families can take an extended vacation, increasing their festivities and rest. American workers need time to spend with their families, grandparents, and loved ones. This RV week will increase time off by seven days, permitting commutes to many national locations. America needs to stimulate economic growth and get people ready for a new life after this COVID-19 pandemic.

Americans on vacation spend money on food, fuel, lodging, recreational activities, and family products, including souvenirs. Many of America's cities depend on tourism along with other businesses encountered by traveling families. Jump-starting America's economy is what this nation needs. Families need to see other parts of the United States affected by this pandemic so they spark new ideas to fix internal problems. Parents on a road trip may decide to move to another state, invest in recovering American businesses, or start a new business, just because of traveling with their children. Parents can teach their children many things if they have more time with them during a road trip. Companies and businesses selling recreational vehicles will make profits as families who lost homes are encouraged to invest in RVs.

In the United States, the American bison is a natural symbol of the nation's indigenous wildlife. There used to be over sixty million bison roaming the United States wild and free. Why hasn't anyone in government ever proposed trying to reestablish sixty million bison in the lands of America? Why is it that some people always take away so much life, but they spend very little time desiring to replenish it? Why hasn't the US Congress enacted a conservationist bill to make all open lands around vast military base perimeters breeding grounds for wild bison to flourish? The US military bases need mascots as symbols of America other than the eagle, which is also endangered. I've seen huge expanses of land in the United States that could be protected by military bases to enable bison to repopulate. Why can't Americans live in peace with nature's creatures, which also need space to reproduce and live in safety? I propose that the US government consider giving Native Americans their lands back to establish peace. The American government needs to start helping animal species that once thrived across America to again thrive across America in vast numbers.

These two proposals would enact positive change in this nation. When Native Americans are given land, the land will be beautified, and animals will be able to find sanctuary within enlarged reservations. The

US government can meet Native Americans halfway by giving them vast amounts of land, which can be used for conservationist purposes to protect wildlife.

To replenish sixty million bison, much land will be required to corral herds and isolate them where they won't be hunted. Can the American government give back more than it has taken over the centuries? We need to immediately clean up all of America and make things right. For when government will not provide justice, then all we have is tyranny placed upon a nation's people.

The US Congress needs to pass a new law that anyone owning animals, including livestock, must provide them shade from the sun's heat. I would like to see netting over pastures, which can be retracted on metal cables or ropes between support poles, to provide shade. The animals would have much-needed shade and increased access to water. The heat from the sun is also completely drying out forests, parching the earth until forest fires ignite.

Animals can get cancer from the sun just like humans but in a different way affecting their flesh, which is consumed. The worst corralling system for animals is not providing them ample space to move around. Allowing animals to lie in their own feces full of bacteria and viruses ready to be sold as meat for human consumption is sickening. How are humans getting all these cancers, gastrointestinal problems, and strange diseases? Feces that come into contact with crops and livestock are catalysts for humans to catch new diseases.

The best farms in America have perimeters on all four sides as wide as ten tree canopies fully grown and evenly spaced. These tree barriers around farms and ranches provide sanctuary for moths, butterflies, and honeybees, providing cross-pollination of crops. These tree barriers also reduce topsoil depletion while adding important organic material to the surrounding soil.

These perimeters around farms also provide shade and sanctuary for livestock on hot days. Other wild animals also use these spaces between farms to raise their young, forage, rest, and find sanctuary in the winter months. American farms need to plant ten tree barriers along their perimeters to help reduce CO_2 in the upper atmosphere. A global initiative can encourage greenhouses and arborists to help all farms and ranches plant these trees.

POLAR BEARS ARE LOSING arctic habitat, so let's tranquilize families of polar bears for relocation to protected wildlife areas. Let's move some families of polar bears with their young to Patagonia to live on a cluster of islands where seals and other wildlife reside. The land surrounding Mount Sarmiento in Patagonia would be a perfect habitat for polar bears. The polar bears will mate and reproduce, increasing their numbers. They can be microchipped or equipped with tracking devices so they can be studied throughout the year. Maybe some polar bears can also be relocated to Antarctica so research teams there can study them.

The fact is, massive wild fires point to a problem that global warming isn't going away. We can fly helicopters in a straight-line formation, carrying metal cables supporting flexible water hoses for siphoning water from lakes and reservoirs. The forests are burning around neighborhoods and cities without being sprayed 360 degrees around to protect them. If humans aren't able to protect their own habitat, they aren't prepared to protect that of the animals.

7

GLOBAL CITIZENSHIP

Why hasn't the United Nations united humanity in global citizenship? United Nations members must partner for global peace to make the world better through an undertaking that cannot be ignored: global citizenship. Technology has brought all nations autonomous vehicles, high-speed trains, hyper tubes, supersonic jets, and ocean vessels transporting citizens rapidly to international destinations.

Global citizenship is not just a travel document like a passport but an ability to partake in each country's union in the United Nations for the prosperity of all citizens. Globally, citizens believe international travel is as normal as using the internet for communication. The earth will have a population of nine billion in the near future. The United Nations has an obligation to support the children of this world who are raised with technology.

Nations share vaccines and medicine with each other; the global citizen is entitled to these privileges. United Nations members need to share the benefits and privileges of their countries with international citizens who abide by global laws. Americans overseas during a pandemic need the same benefits and health care as the citizens of the countries where they're currently residing for their survival.

When each country shares equally in a health-care system, everyone benefits from the transparency. International students need health care and privileges like everyone else they share this planet with. Global citizens are developing businesses, purchasing homes, living abroad, acquiring retirement housing, securing health care, and seeking advanced education. Global citizenship reduces the risk of being harmed during international travel. All global citizens seek prosperity to live better lives, aiding their families as well as themselves. We need global unity because, during chaotic events, we rely on each other for survival.

The United Nations has a responsibility to unite all the countries of the world and to enhance prosperity for the global community. All global citizens need their human rights improved. When honest nations work together, human trafficking, crime, terrorism, and internal corruption are all reduced.

The Geneva conventions need improved United Nations oversight. The United Nations needs to overhaul global health-care standards to enhance human rights for all. Refugees have arrived at numerous borders needing to share in each country's benefits for humanity to bring forth peace.

Establishing refugee camps that can provide a service is beneficial to all seeking prosperity. There should be agricultural programs for all refugees in every nation. Helping refugees erect sanitation facilities is good for the environment, and the instruction provides a lifelong skill.

When refugees can build communities to support themselves, this reduces any burden they may create on a nation's economy, which has to allocate funds to feed its citizens. When we care for those who are ill, we prevent others from becoming ill. When we feed those who are hungry, we prevent ourselves from becoming hungry because people learn the value of creating farms. When we help others overcome hunger, they become stronger, and then they, in return, can help us with the labor for the harvest. Refugees need jobs, and all nations need crops to help eliminate world hunger. The United Nations must address the needs of the global population.

World leaders should endorse ten new global rights for all humanity. Global citizenship grants equality for each citizen belonging to the member states participating in free trade for prosperity. The United Nations is responsible for facilitating unity among the different nations.

Global peace is an objective for all member states. The United

Nations' responsibility to humanity is to enforce a standard of care for all citizens of the earth to eliminate malnutrition and hunger. The United Nations cannot neglect increases in our world's population. Increases in population will create environmental problems, global food shortages, conflict, health emergencies, human migration, and wars.

Addressing these concerns now will reduce global problems in the near future. All citizens of all races will share in this global citizenship free from violence. We now find nations sharing medical supplies during a worldwide pandemic. Creating global citizenship with all nations helps improve the economy of the entire planet.

Having citizenship in a hundred countries is better than having citizenship in just a few countries. The benefits of belonging to the United Nations' global citizenship program are astronomical. We must promote the acceptance of humanity's cultural diversity by unifying humanity in mutual cooperation.

People will become less racist as they experience new cultures, eat new foods, and partake in different forms of business internationally. All nations seek to make life better for their citizens. When we unite people from around the world for positive change, great things can happen. Children from all nations have an interest in outer space technology, and when we unite them to pursue living off this world, then great progress can be made.

Global citizenship will bring brilliant minds from many countries together for progress instead of driving counties apart and against each other in war. Global citizens will be able to partake in shared meals with international families to bring about world peace by uniting us to reduce terrorism. This generation will be living in a world with ten billion people crossing borders on mass transit as a normal way of life.

This generation will become accustomed to migration from one country to the next for business projects, family enhancement, and cultural advancement. The United Nations must facilitate a new structure that creates benefits for global citizens. Enhancing productivity in all nations is one United Nations concern that must be overcome by implementing new trade opportunities.

Members of the global citizenship program will be able to exclude nations with ties to drug distribution, terrorism, human trafficking, sex crimes, and major fraud, for example. If a nation wants to be included in the global citizenship program, it must prove that it's making internal changes to eliminate corruption. A standard must be met to obtain the privileges sought.

When we start working together instead of being divided, our nations will become prosperous with more organization, healthier communities, and a cleaner environment. Today, global citizens often affiliate themselves with more than one race and/or religious belief. The United Nations has a responsibility to make global citizenship available in all member states, guaranteeing everyone on Earth an equal opportunity for progress.

The United Nations must offer global citizenship as a means of creating unity for all of humanity. Global citizens will be screened and will be entitled to travel discounts. Boosting tourism in every country in the global citizenship program will be beneficial for all member nations. Economies will be stimulated as more people travel, and poor countries that belong to the citizenship program will receive more business than nations not belonging to the initiative. There can also be benefits for nations that accept entry into the program early rather than opting to enter it late. Those opting in later may have to pay a higher fee to belong.

Funds collected for global citizenship will help pay for the enhancements made in each country choosing to belong. Nations that work hard and contribute to the global fund for improvements will receive more benefits by tying themselves to other nations with resources, equipment, technologies, and industry. The United Nations must fund new sanitation and sewer treatment facilities, which aid the public health of massive populations. The United Nations must fund hospitals, clinics, recycling programs, housing development, farms, fisheries, reforestation programs, prisons, and environmental protection cleanup programs.

Every country needs a new way to rebound from economic problems. Increasing trade and tourism helps businesses struggling to endure during a difficult time in the economy. Only the strongest will survive in the business world. When we unite the entire planet in global citizenship, many corporate conglomerates will rise to be pinnacles of leadership. All members of the United Nations need to do their share of the work to correct global economic infrastructures. Solid leadership is what the world currently needs to make progress for an environmental change in a new direction.

8

CHANGE STARTS FROM WITHIN US

The official job of the president of the United States is to protect American citizens from any harm, foreign or domestic. As president, I will make it my duty to stir many emotions in America if it guarantees Americans greater safety in these times of violence. Harm to citizens is a major problem in the United States, and more criminals need to be imprisoned. Decent citizens need to be enriched with prosperity that's free from oppression. Devious acts intended to deprive citizens of their hands' hard work will be met with swift punishment.

Nations are built on foundations of integrity with laboring hands being honest and truthful and attempting to do good. I hope Americans will vote for me even though some of my statements are controversial. When I'm upset with a problem, the American people will have me honestly addressing the problem and disclosing my decision with sound reasoning. Rather than speaking about past presidents who may have been deceptive, I'd rather be an honest president and make the problem known to our nation. Presidents must obtain congressional approval to address international problems affecting Americans because all these decisions have global repercussions.

Every administration will encounter times when criminals try to thwart the efforts of the presidential cabinet. I'd like to assure Americans that during my term in office, I would emphasize maintaining peace to solve our international problems. Going to war never seems to fix problems completely. As long as Americans aren't attacked, there's no need to destroy years of peaceful progress with retaliatory air strikes that stir up strife. I will not tolerate terrorism during my term in office. Establishing a very well-educated cabinet with brilliant university minds early on during my term in office will be a priority. Americans cannot fix all the problems other nations possess. It's time for other nations to become more responsible for their own internal problems. Each nation in a healthcare crisis has the duty to provide their people all necessities for survival including proper nutrition. United global agriculture, aquaculture, horticulture, and habitat restoration for zoological species must be our emphasis for our growing human population.

I will strive to make America prosperous again. I'll fix America's internal problems by revising government documents, making sure funds are efficiently spent where most needed for the betterment of our nation's communities. A strong educational system is the core of fixing all America's problems. Don't let opportunities to invest in our humanity go by the wayside. Help me become the president of the United States. I'll do my work as efficiently as possible.

Being able to work with others in Washington, DC, is the cornerstone to crucial successes for our nation's future. Rather than dividing the various political positions held in Washington, DC, I'll make it my objective to fix American infrastructure by working closely with politicians. Politics in America today needs to go back to its roots, requiring us to work harder than our neighbors to reap the rewards of a good day's labor.

Hard work is where progress is made, and America needs to fix what we already have before making anything lavishly new. I'm all for new cities being built, but let's first fix the cities that have been contaminating our nation's landscape. Let's clean up America on a massive scale. Let's start a program in which truck drivers press a button to mark any GPS coordinate where a pothole exists on our roadways, and let's fix potholes within forty-eight hours.

We have the technology; robotics; hydraulics; manpower; vehicles; and newer, stronger construction materials, so now all we need is the initiative to get the job done. Stopping accidents on our nation's roads will save countless lives and help our transportation system's efficiency when we need crucial supplies delivered safely. Rapid processes getting roadway repairs done quickly shouldn't be hard for our young, innovative minds with brilliant robotic designs.

During our nation's pandemic, electric vehicles would have been very useful to deliver meals to the millions of residents of our cities. Green-friendly electric vehicles have low maintenance requirements, needing only tire replacement. Delivery employees racked up many millions of miles on their vehicles, increasing their burden while providing so many families an invaluable service. We must make improvements as a nation and find better ways to aid each other during difficult times.

I'd like to see our postal service becoming green friendly beside our nation's school systems, by operating electrical vehicles. Teachers chaperoning children on electrical buses to all our national parks, museums, planetariums, universities, biospheres, and scientific facilities for learning purposes would be progress. Students shouldn't be kept in classrooms and deprived of the joy of experiencing life interactively. Students need to see and touch our nation's achievements.

As president of the United States, I will make it mandatory for the educational system to facilitate travel programs locally, statewide, and nationally as part of the curriculum. Students from all over the nation should visit the National Mall and tour the Capitol, the Library of Congress, and the US Patent and Trademark Office as part of their week or two-week-long excursion to Washington, DC.

Students have witnessed the chaos between Republicans and Democrats opposing each other. The politics of America's founders are not the politics of today. The elephants and donkeys are the political parties of the era before. I'm sorry, but grandfather's analog design is no longer applicable today as we move forward progressively with digitized technologies for space exploration. We're living in a very technologically savvy generation with young students needing to establish their own changes to American policy.

Americans don't want to see more donkeys and elephants were ready for the political groups like the Clydesdale Stallions, Bison, and the Blue Whales, for example. America has all these great creatures as part of its national heritage residing with the farmers, Native Americans, and the sailors navigating our planet's oceans. We need political groups that will focus more on the people's needs: our environmental issues with the Blue Whales party, preservationists of heritage with the Bison party, and the conservationist farmers our Clydesdale Stallions party.

Three parties divide power nationally, and the fourth, much smaller political power is the Eagle party. The Eagle party is responsible for our off-world management of space outposts and bases, including our colonization of Mars. The Eagle party is a political structure designed in the event a major cataclysmic event occurs on Earth, destroying it and leaving humanity to survive alone in outer space. The Eagle party is a continuity of government contingency ark plan for starting over should our Earth be destroyed.

The Republicans and Democrats are antiquated eras for progress having created trillions of dollars of debt. The US Constitution needs new modifications which were not foreseen by the past generations living hundreds of years ago not envisioning technologies advancements. We need new parties run by this generation with new mascots unionizing American politics. The private clubs and mansions of the past, which were once nice, have gotten cobwebs, becoming like mildewed second-class estates. This generation of young people needs to erect their own clubhouses with political groups to phase out the Republicans and Democrats hindering progress. The Republicans and the Democrats with all their followers aren't the most intelligent groups it's the international valedictorians who are the hard working meritorious achievers for success.

Political cartoons of stallions stepping over donkeys, blue whales swatting elephants to fly, and bison overwhelming their opposition by their sheer number must come forth. The landscape of American politics needs a revision because the past way of doing things isn't as effective. Globally, nations have witnessed in the media just how imbalanced democracy is. Our nation's capitol is a global symbol that's been trampled on, all because huge problems remain unresolved. On your parents political watch, America has had its national systems and security designs compromised by the Republicans and the Democrats.

America cannot afford to forget others, because forgetting others has gotten us in the mess we're currently in with division in this nation on many political issues. Forgetting others cannot continue much further without each branch of the federal government going through some changes. The strongest nations will not forget their people and will not treat their people as subhuman. Changes in America's three federal government branches cannot be just external, they must be internal as well. A government not fit to lead any people allows wickedness to happen to its people while refusing to punish tyrannical wicked doers that are being criminals.

As president, I will work to make an educational travel and learning experience open to all students throughout our nation. Parents who feel more comfortable traveling with their children will be encouraged to participate as chaperones. Our nation's students are the stewards of our country's future. We will work with all lodging facilities, air bases, airlines, rail lines, and bus services to arrange transportation for all students to visit our nation's capital. Students will visit national landmarks as a mandatory interactive history experience.

Let's bring Americans together and stop all the school violence with peaceful, interactive activities for all. Students will have the opportunity to write a report on their experiences and to continue their studies in the classroom after their participation. All students will be fed meals during these excursions, so no student will go hungry while other students eat. The pandemic has kept students inside and unable to enjoy the fresh air in our nation's parks. As president, I'll open up our beautiful national parks for students to enjoy the natural environment for a greater appreciation of what Americans have.

Traveling should be a biweekly experience for all American students, even if it's only to local establishments, companies, air bases, museums, or science centers. Teachers who need a break from grading papers can interact with students in educational field trips requiring all students to perfect their manners as Americans. Students need to be taught how to behave as part of their curriculum because being respectful is a lifelong value that conveys how they should treat others, including their fellow students.

We must stop the school shootings, assaults, and bullying taking place in America. Being respectful, kind, and upright will become part of the American educational system. American students will begin improving their socialization skills immediately, and that's a very important part of any educational advancement. Proper etiquette is a lifelong skill that all students need for any occupation while working with others. Students will be sent on many field trips, and they'll be required to be respectful to all people whom they encounter.

If students are not taught proper behavior in their homes, we will teach them in our American educational system. In no classroom nationwide will any improper behavior be tolerated from the start of elementary school through high school. Students will begin an early education in how to treat teachers, parents, and professionals with respect. Students must learn how to be respectful to the members of government they may meet in their visits to Washington, DC. Our teachers cannot feel threatened in their workplace, and the name of any student affiliated with a gang will be recorded in a police file immediately. Students will be made aware of their police files to thwart bad behavior. Our communities need to change because we're a nation of diversity. We need our students going out into our communities and cleaning up the trash as part of learning about their environment. We have too many cities, highways, and parks becoming homeless encampments and creating more waste in our communities. The sooner we teach our nation's students the importance of cleaning up their environment, the better America will become as a whole. I propose the Beautifying America (BA) program for all schools.

We need to make America's roadways safer for all travelers by rapidly repairing the damaged pavement to reduce accidents harming our nation's people. Lets create eight hundred jobs with four teams of four people in each team traveling within their state marking in orange paint all potholes causing depressions on roadways and bridges? Recreational vehicles can tow a service vehicle and the road teams can live on the road for a one year contract in all fifty states. Road teams travel all the roadways in their state daily marking hundreds of potholes as a federal government oversight aiding commercial trucking to get the repairs done. Once the potholes are marked in state road crews must fix the potholes within a forty-eight hour period. Let's get Americans working even if we can't fix all the roadways lets fix the most important problem sites with potholes. Traveling pothole marking crews will receive fuel cards, meal stipends, and a decent salary to help improve America's infrastructure.

I'm sure my words are controversial to many who will disagree with me. I will continue to work on constructive projects despite the efforts of some who've chosen to hated me and wanted to prevent me from learning or achieving success. I will continue to invent, making contributions to science, medicine, reforestation, marine biology, engineering, aerospace, astronautics, mathematics, and the arts. I'm confident that the readers of this book are making plans on how they can contribute to the advancement of humanity. My advice is watch out for those who want to stop you from reaching your maximum potential in life because often it's those people you least expect who see your gift.

IF I WAS to build the world's tallest megastructure, wouldn't I do all the mathematical computations to ensure the building's structural integrity and longevity? Wouldn't I analyze past and current building failures to mitigate the possibility of making a mistake? Wouldn't I also spend days, weeks, months, and years putting down every detail on paper for evaluation or revision to make the building project a success? Before any building can be built, a cost assessment must be done for the construction, and who's going to pay for it? Will the project be kept secret long enough so that the best deals can be made to cover its costs?

The privacy of an inventor's infrastructure plans in civil engineering, architecture, and business are for those of us who stand in corporate offices, not for those uninvited to these private meetings. Beware of those who think they're entitled to breach your privacy because of their age, degree level, former rank, or current rank. Many businesses are private, so the mindset that this data is for others to know or access is criminal, and compromises corporations.

BEWARE of mentally ill narcissists who think that everything you're working on is something that they must get access to. Some people understand privacy, but they just don't want others to have it, for then they're not the center of attention. Delusional minds want to exalt themselves like they're smarter than everyone else. I've been the victim of more than one narcissist stealing my identity information and private intellectual property documents while disseminating stolen information to a network of criminal homosexuals like themselves.

I've spent my lifetime gathering detailed information for my projects while the narcissists smiled to themselves, thinking that they would steal everything I was working on when I was done. It's no different behavior in government, for others want what they do not deserve. Attacking the souls of people isn't normal, and someday I'm writing a book on mental illness to give insight into the dark minds of delusional people. Beware of false prophets—the people who are convincing but are conscienceless, who strive to manipulate others into believing their lies.

Everything that delusional mentally ill people harboring their insanity show you about themselves is a total lie, including their possessions to their clout and knowledge. These narcissists have lived their entire lives being two faced, only showing you what they want you to see and convincing you to believe it. Evil has an expiration date, for the truth is made known in the light, and it's all recorded in heaven. Delusional people with wicked hearts persist in thinking they can hide what they've been doing because they think no one will doubt them or their authority.

A mentally ill person will give a hug or kiss to the very person they've secretly been poisoning, while being seen by others as being compassionate. To protect yourself you must eliminate the dangerous person indefinitely from your life, especially if they're toxic, using illegal drugs combined with prescription drugs and alcohol. So many people have become victims because they let down their guard, thinking the other person would not snap, and they did.

Beware of deviant family members living predatorily with mental illness who strategize deceptive ways in their diseased minds, making human beings pawns in their game. Delusional minds think of and create plans on how to be harmful towards victims through their ways of inflicting torment upon souls repetitively. People often wonder why I repeat certain statements in similar or slightly different ways. People who were the victims of serial killers, mass shooting incidents, sexual assaults, or imprisoned in dungeons never thought they would get themselves in such a situational circumstance. If I can place emphasis on saying something frequently this maybe exactly what saves your life from a dangerous situation. Recognize the problem early in another person see it, and know it, because your smart to know something just isn't right so go with your intuition. Feelings are so important and blocking out what you feel sometimes just by being overly nice is like placing a target on yourself to someone who just doesn't care but only wants. The brain of deranged people are missing many important parts especially the conscience it's absolutely absent one hundred percent non existent. How can a human being cannibalize another human being and think nothing of it? How can someone participate in theft causing tens of thousands human deaths because they warred against scientists researching cures for diseases or preventing medications from being developed? Demonic people don't care about anything except what they want and thinking their mind is greater than anyone else's mind or thoughts making them very dangerous personalities. Women who cannot escape being abused in domestic violence need to get far away from the diseased mind. The same can be said of siblings dealing with addicts who are out of their minds living crazy on illegal drugs. I say yes in these days with violent events happening frequently a protective firearm is good to protect your children including yourself from psychotic predators. Protecting ourselves from the insane people is part of what the US Constitution granted us so never let a politician take away your ability to save your own life. Politicians have firearms and resources to protect themselves so don't let judges, police, federal agencies or unjust people serve you up to be slaughtered by psychotic predators.

I point out details about mental illness in different ways, and this is done to give you different forms of protection from those who would seek to do you harm. A mentally ill person will put so much energy into violating a victim's soul. Offending the hearts and souls of others is like a daily cuisine for narcissists, for they thrive on attacking the holy spirit which dwells within those they victimize. Mentally ill people think they're indefinitely above others in every capacity of knowledge, and this also includes the religious beliefs people hold sacred.

Mentally ill people think for all eternity they're indefinitely more important and more favored than you and that your insignificant to them especially if they've ever served in the military.

Many deranged and delusional psychotic serial killers have come forth out from the United States military so don't be so quick to green light the minds of these groups of men or women who live clandestine lifestyles. Some veterans from the military are fine mentally but there are sadly many veterans who aren't fine mentally who really are delusional cognitively. Be cautious of veterans getting involved in churches this can be a smoke screen for the lifestyle they live once the doors close at home or where they go out of view. Frequently veterans will aid other veterans in their games of deception thus creating a entanglement for many sins to ensnare minds for murder, illegal drugs, sex crimes, and criminal activities. The Holy Bible teaches us there are rewards for the faithful who will receive crowns for being saints. Beware, for anyone can have mental illness, including members of the military, police, and even those holding government occupations. Narcissists think they supersede you indefinitely in all things. Narcissistic personalities will find it very comical to offend you, especially if they've stolen a huge portion of your life as well as sensitive information.

The greater the violation, the greater the reward to bring joy into the narcissist's life striving to offend you. Delusional minds live in an ongoing fantasy with one objective: to continue undermining your soul's depth in every facet indefinitely. Mentally ill people who formally held any rank have sinned worse than those they judge and want continual dominion over. Beware of minds that can never live within their own perimeters but invade your space.

When a mentally ill person commits a mass shooting, there's no cost savings for the harmed families, lawsuits, court fees, and massive damage done ruining so many lives. Locking up mentally ill people is a cost savings in the long run. We need to funnel massive amounts of money into megastructure mental health therapy facilities designed to be like resorts with all sorts of programs. Mentally ill delusional people do not spend their days seeking Yahweh; they spend their days thinking up a strategy to undermine others they want something from.

Mentally ill people find pleasure in doing illegal drugs and consuming massive amounts of alcoholic beverages with like-minded addicts. In fact, religion is a perfect platform for a mentally ill person to be deceptive. Mental ill people will live as a pillar of lies while orchestrating a greater deception, masquerading among congregations or choirs being holy. Satan the Devil was also deceptive, masquerading among choirs being full of sin.

The biggest mistake society has made is trying to teach mentally ill people about religion. Stopping the use of drugs for a day, week, month, or year doesn't provide enough time to rewire a brain to understand religion. A mentally ill person's mind will go off course into strange thoughts when being fed religion. Many mentally ill people find drugs and still use them, even while in rehabilitation programs.

Minds that are influenced by numerous illegal chemicals rewiring their brain's circuitry don't need to be fed religion when they haven't fully recovered in these short rehabilitation programs. Religion is best kept to a time after a person is 100 percent off drugs, keeping their minds from wandering astray in delusions. Getting someone off drugs may take decades—just look at the homeless encampments nationwide and around the world.

As delusional as it is, mentally ill people doing drugs and abusing alcohol believe they deserve a penthouse. Mentally ill people believe they deserve to have access to everything a penthouse owner owns or is involved in. Sometimes security or staff in hotels and high-rise buildings discover mentally ill people sleeping outside doorways to rooms or residences.

The real danger is when new narcotics hit the market and are made very inexpensive, even for a beggar. The damage to the brains of homeless people can turn someone considered calm or harmless into a raging psychotic. If the illegal drug is consumed while the homeless person is already on a building's upper floors, guests or tenants are in a great danger.

The date rape drug Rohypnol may find its territory taken over by the Devil's Breath (DB), a chemical compound cartels are experimenting with in different combinations. Mentally ill people who own a home, condo, or apartment or live under bridges have spent much of their lives inhaling different chemicals as assorted drugs. Many of these illegal chemicals in drugs induce thoughts of grandeur. As delusional as it is, mentally ill people who are thieves may find the Devil's Breath a perfect fit for their lives of stealing.

Using a chemical to put victims in a zombielike state could be the future of America, with more people committing robberies, finding new ways to acquire wealth. Some mentally ill people have been in Veterans Administration psychiatric departments, and they think they're smarter than doctors living in elite neighborhoods or high-rise buildings.

We live in an illegal drug society with so many people having an abnormality of delusional thoughts. The minds of mentally ill people believe they're entitled to all the things their hands never worked for. Hollywood has seen mentally ill people come and go, believing that they're the most deserving of the highest paychecks. What goes through the mind of a mentally ill person who does not get what they want? Jealousy can fuel many mentally ill people to develop a knock, knock, pounding-on-the-door, home invasion mentality of revenge.

Mentally ill people steal and think that their hearts possess a deeper creativity than anyone else around them. When the mentally ill person is homosexual member of the LGBTQIA community, often they'll participate in the similar types of delusional activities with others in their network repeating harm to others with varying degrees of severity. Children aren't the only victims; adults can also be the victims when poisoned or illegally drugged.

I highly doubt a person who smokes marijuana and drinks alcohol on a regular basis is going to spend the time designing the world's tallest mega-city or developing Hollywood's next big hit movie. I have a reason to be upset, for the crimes committed against me were intentional to harm my financial infrastructure, ruining every aspect of my life aiming to help humanitarian projects. Jesus Christ and his disciples were stolen from also by a thief striving to ruin their plans having a heart closer to demons rather than closer to Christ. Often the most devious of sinners is the one that people least expect who has assets, clout, rank, and a tongue that can sway the most intelligent minds to be in their favor. Be confident that the sneaky deceitfully evil people who wickedly think they can get away with the crimes they minimize are the ones most despised by our Lord Jesus Christ.

Getting over others younger than myself who now have greater success than me has been difficult for me because I feel robbed of the time I put into figuring out many things in this world. I've achieved very little success that I originally sought to achieve, and it wasn't that I didn't do the work; I was victimized repetitively by narcissistic thieves. The thief stole my civil engineering plans, power distribution designs, architecture plans, medical designs, and an extensive portfolio collection of all my inventions I was going to patent. Working on projects over and over again each day is your time and not someone else's for it's your improvements in your artistic skill. Separate yourself from the critics because many people without talent will be quick to be critical of your brilliant ideas either being condescending of them or quick to steal what they want. Some of the most valuable ideas, art, writing, or pieces of property in world history were rejected by many critics before becoming multi billion dollar franchises. Don't allow your family members to have their way against your soul. My experience is wicked thieves do not stop, they only lie and say they'll stop! Family will cling to their sinful tangent which is nigh onto the demons who also cling very near to those committing sins against your spirit. How do demons find their way into households and people wonder why so many strange things are happening? Demons find their way into households because the mind of a person open to accept they're superior to another person while living in sin is just what demons need to possess a prideful mind.

My advice beware of family members who want to judge you when you know your own soul has greater talents than them and they want to steal what you have by sabotaging your soul's goals through theft. If you know you can win an Olympic medal then get as far away from them as possible because they're the doubting unbelievers you don't need warring against your soul's time.

I spent my days studying, writing, drafting, researching, designing new architecture, conceiving presentation models, and even talking with professionals who built skyscrapers. All my life up to the age of forty-five was stolen by delusional members of my family stealing from me repetitively. Striving to make my face invisible has become my siblings indefinite tangent to fulfill their hatred of me. The police have allowed me to be victimized and even sided with those doing harm to my life even after I went to file complaints. Remember even Jesus Christ was made a victim under the authority of those unworthy to judge him so when you encounter this hatred think only about heaven's treasures. The treasures of heaven no one can take from you for Yahweh knows who prayed, who confessed their sins, and who sought to better themselves.

Family can be against you enriching themselves through sin but Yahweh is one hundred percent for you to be richer in heaven. What's the most disturbing part of having my soul attacked is that each of my family members participated in the vexing of my spirit, striving to steal my soul's life. When something is wrong with another person's brain, we must expose them in our society to stop the madness. Evil hearts will continue beating until we stop them from continuing their delusion. We need more mental health facilities and Catholic exorcisms for those who possess wicked spirits within themselves. I sadly have to say that some people prefer evil and reject Yahweh's guidance as well as the churches teachings to the fullest extent. Not everyone can make it into heaven but when you aim your soul on Yahweh and Jesus Christ you will find the path to ascension many fail to feel or wield throughout their bodies transforming to be holy.

We live in a society in which people steal from cash registers, bank accounts, purses, wallets, phones, computers, corporations, the government, and even social media platforms. Millions of pennies missing—who will notice? Thousands of written entertainment ideas missing—who will notice? You may not even know that you were stolen from because your soul is being undermined, and it happens often because criminals want what you have.

Stealing to make people forget what they have and targeting people who are more prone to forgetting, like patients with injuries or the elderly—who will do that crime but a mentally ill person? The drug Devil's Breath can induce a state of amnesia, and there are thousands of incidents of people using this on their victims. There are people with mental illness who have abused elderly seniors, children, and families. There have been elderly women who've been raped and children molested because of Devil's Breath.

I'm giving you a warning that these people who are out of their minds aren't going to stop their ways. Addicts will continue to be delusional and deceptive until the day they snap because they've fried their brains on psychotic chemicals. Addicts will live like they're on a roller coaster going up and down, enjoying all the strange feelings until they achieve an unexpected psychotic episode. Be careful: delusional people will want to mess with your mind to get you in trouble with law enforcement by tampering excessively with your spirit behind the scenes to stir up strife.

A mentally ill person with a vendetta will want to see you imprisoned and losing your life while they covered up all the ways they disturbed your thoughts, beliefs, and emotions. Getting the help of professionals like psychiatrists and police departments to evaluate the person disturbing you is invaluable. Professionals will aid you in a plan of protection from such threats very near to you.

Getting far away from the diseased mind is a mandatory avenue toward making the right choices for your well-being. Narcissism dwells also in heterosexuals and bisexuals, so don't dismiss family members with alternate lifestyles who can be very toxic with their thoughts.

LATINOS WILL COMPLAIN when bad times happen, but in good times, Latinos refuse to invest with the right people who are Latino. Latinos cannot have it both ways. Either they invest in their own people, or they do not. There's no room for complaining that there aren't more Latino-owned businesses, skyscrapers, or companies when our people refuse to invest in our hard workers.

Every day, great business opportunities are lost so that the prosperity goes to others, rather than to our own people. Change starts by investing in those who actually did something visible and not in those who are merely well spoken, having an ability to sway minds. The world has too many con artists working in every occupation, seeking out new ways to skim profits.

It would be nice if Spanish-speaking people around the world all contributed to one megatall skyscraper project I would like to build. I would like to create a scale model of a city and then have a video showing what it would be like to live inside the city. I believe that any new city would need a Vatican presence to help the people living in it to be fruitfully at peace.

We need a new, modern city to be built, employing many laborers who possess an aptitude for artistic work. If all the Catholics worldwide contributed to a mega-city being built, it would only take two years to start the project after financing. Any well-thought-out design must have dedicated architects and numerous revisions for civil engineering of the utilities. On the watch of the Republicans and Democrats bickering amongst themselves there's been breaches in our national security. The Republicans and the Democrats are becoming like antiquities in the vision of our young students who are valedictorians. From all fifty states there needs to be a movement for thousands of brilliant valedictorians having a high education to begin making their political move to replace the Republicans and the Democrats. New political parties which make our environment the number one priority to ensure the continuation of humanity is necessary. Yes industries need to continue developing and being innovative but we need to make progress smartly working together on a global scale to ensure we harm our planet no more through pollution or mismanagement. This generation of young bright new politicians cannot allow the Republicans and Democrats to ruin our economy and perpetuate this nation further into financial crisis with massive debt. All the seats in Washington, DC need to be filled with new minds, and fresh ideas of brilliance bringing us talent able to make the necessary decisions for the best interests of all nations. We are a global community of nations all needing to work together for progress. Bringing forth the changes this nation needs to work with other countries around the world must become a consensus effort for all Americans to establish peace. We cannot tell other nations to be peaceful when Americans cannot lead by example maintaining peace on a constant tangent. Now is the time for students attending universities, colleges, and technical trade schools who will never cease their tangent of collegiate life learning to embark into politics. The 4.0 standards of the past education system are like a 2.0 equivalent in our modern educational system. More and more young people have earned professional degrees so it's time that we become more professional with higher standards for excellence in our American educational system. The straight "A" students with a passion for progress we need to get into the political seats of the Senate, Congress, and all agencies within this nation's Capitol. These valedictorian students have the leadership skills to run this nation with a more effective strategy than our past leaders lacking the vision they haven't attained. Many Americans believe too many political seats have been stolen and not earned. Doing the hard work is what many have failed to do in Washington, DC leaving out those who did the hard work and were never credited for their focus or determination in completing important tasks. The issues in American politics derive from many who don't have the intelligence level to have the positions they do being the thieves who stole data they should have never possessed. Political chaos throughout America is an internal factor rooted in greed rather than a true desire to do the best job possible. So many in Washington, DC want the titles and badges so now it's the time for the true valedictorians to root them out by testing every person in the nation's Capitol through a replacement. Since a Presidency can be compromised the replacing of a nation's leadership with young minds who could have had no involvement in criminal activities is the way to improve national security. Being efficient means creating trillions of dollars of surplus money building banking interest versus having trillions of dollars of debt. The future of progress across America starts with uprooting the past political infrastructures of this nation and summoning multi billionaire families holding this nation together asking them for their advice on fixing real problems.

One life goal I have toward the later end of my life is to run for the presidency of the United States. I'm confident I will immediately encounter opposition minimizing my intelligence for such a position as the presidency. There isn't a subject I cannot learn. When we aren't experts, we delegate important duties to the surgeons in their fields to conduct the upmost professionalism in every occupation. Presidential cabinets are structured with teams working together, not just one person providing a decision but a consensus of educated minds utilizing advisors. I'm not concerned about not being elected to the presidency of the United States because I'm confident in my skill set. I will bring awareness to many of America's internal problems throughout my campaign with the intent to fix many of this nation's obvious problems.

I'm campaigning for peace, space exploration, and the physical as well as medical needs of humanity. I support the global denuclearization —of all nuclear weapons, which are the most unintelligent things that could ever have been built. Nuclear weapons will destroy the planet we all live on, so these weapons have no place in the advancement of civilizations. Nuclear weapons may only be useful for protecting the earth from asteroids, meteorites, or other threats to our planet. I support removing nuclear warheads and putting them underground within sealed-off bunkers.

Components of nuclear missiles can be repurposed for the advancement of space technology. Repurposed rocket components can increase the number of mission launches to other worlds within our solar system for colonization. Denuclearized rockets will enable more supplies to be sent to Jupiter's moons, Mars, and distant worlds for space colonies. We must start new construction developments in other worlds, even if the undertaking is difficult. America needs more educational programs for science, technology, engineering, and mathematics competitions offered to students from many nations. Peaceful competition is good humanity striving to advance all nations.

When international students are having fun and getting along peacefully in wholesome scholastic-based entertainment, we all benefit from politicians fighting less over trivial issues. The sharing of information and the encouragement to labor for sustained living is the direction all nations need to take. World populations are growing, and pandemics are becoming more frequent occurrences, disturbing our society. Doing construction in other worlds should be the mission for this generation of students experiencing global warming, natural disasters, pandemics, and agricultural shortages. We've all had to learn to live in confined spaces, and now students should apply this to living on Mars or even Jupiter's moons. There will be no taxes for people choosing to relocate to Mars for colonization. It's the parents, teachers, and politicians who need to unite in motivating students internationally for a mission to Mars movement. We need students from all countries inventing and designing for

Mars habitation. Students must become motivated for studies in medicine, biology, earth sciences, new technologies, space engineering, and astronomical navigational mathematics.

We need to move forward, cutting global defense budgets by 50 percent and redirecting these funds to space colonization programs. Humanity is destroying the earth, and once it's dead, the only survivors will be those we sent to other planets in our solar system for terraforming operations. Not many human beings will be able to live in a world without vegetation and with all the diversity of other life-forms dead. The people who are able to live underground for a period of time here on Earth will eventually run out of supplies and succumb to the perilous death awaiting them outside their hidden chambers. It's not an easy task living in a world with medicine shortages, lack of food, massive depression among survivors, and the human element of desperation generating conflict. Moving to another planet and building an infrastructure there is the strategy we must employ for the continuance of a human, biological, and zoological ark of life. So much money is wasted on destroying life, and the richest people in the end will be those who possess the greatest number of living organisms.

Human beings have used nuclear weapons in the past; do you doubt that they will use them again in the future? If you do doubt that human beings will use nuclear weapons, I tell you I've seen glimpses of the future revealed to me, and they use them again. It's very important to plan for the future because living day to day, barely surviving, isn't the way Americans should be living. Our civilizations need improvements in fertilizers to aid in growing crops to feed domestic livestock that feed human populations.

I'M ALSO ADVOCATING for the environmental protection of our Earth through a global recycling program. In the global recycling program, many massive prisons will be developed to house criminals. The drug dealers, drug lords, and criminals will labor, doing recycling work for all the years they've harmed our societies with their corruption.

These men who commit crimes will not stop unless they're arrested and locked away for the rest of their lives. The Southern states of the

United States are perfect places to build massive prisons. Prisons are much less expensive than stealth bombers, which accomplish very little for the advancement of humanity. Prisons help in the incarceration of bad men, hindering their murders and the spread of their drugs causing overdoses. Prisons reduce the ability for criminals to conduct thefts, sexual assaults, and other crimes in peaceful communities.

Massive prisons are the deterrent for criminals. We need to go on the offensive to round up criminals to do hard labor, working them in recycling, sanitation, and composting facilities. Yes, society will still see murders and violence, but there will be fewer murders if we lock up all the criminals. Mothers with poor parental skills must also be locked up to stop the overflow of newborns being introduced into the underworld of corruption.

Tens of thousands of murders happen every year, and the criminals continue murdering to send a fearful message to communities. These fearful messages from criminals need to stop. The police need to send a more fearful message to criminals that they'll never stop. Criminals murder because they want their way, so we must wage war against them even harder to destroy every infrastructure they possess so they cannot have their way.

Strange women have been helping hold the empires of these criminals together by being deceptive themselves. Women can quickly say they were the victims and were threatened by drug-dealing boyfriends, but that's just a manipulation tactic. Many women have squirreled away dirty money, drugs, and weapons entrusted to them by drug-dealing boyfriends, for example. The police can arrest male criminals, but if the women are willing to reproduce more criminal men, this revolution means more women need to be locked up also.

NATIONS NEED INTERNATIONALLY STAFFED mega-hospitals with no waiting lists. Patients need immediate services and not delays in their health care. The sooner patients are treated, the sooner they can get back to work. Patients shouldn't have to wait for an American specialist to become available when so many international specialists are immediately available. Around the perimeter of the mega-hospital will be a mega-city, providing housing for employees, patients, retired seniors, and the elderly.

All mega-hospitals will have advanced energy independence, having their own water, sewer, and air purification systems, reducing the environmental impact on city sanitation systems. Health care is directly affected by the spread of germs, bacteria, viruses, and various diseases that make millions of people ill annually, while sanitation infrastructures are often overlooked. Advancing sanitation designs is beneficial to all people and our environment, and we need more inventors tackling these problematic issues. Backup systems are crucial in any infrastructure, even when we consider weather or unforeseen disasters.

We'll start a global program in which a student who chooses health care as a profession will owe no money at the end of their studies. Societies will award free education to students working toward a health-care profession. Societies will, in return, receive the benefit of health-care professionals providing significantly reduced prices for medical and dental services for patients. In my program, medical and dental students will have no debt after graduation and, therefore, no stress or depressing financial worries.

The only requirement for medical students is that they perform a term of service to their countries for the free education they received. Students have to work like everyone else who was laboring while the students were studying. In these uncertain times, with pandemics flourishing, medical and dental services provided to the public are invaluable. We shouldn't burden caregivers with debt but enable them to do their jobs with less stress and no worries during an international crisis. Society will continue to function in all its diversity, even while students aren't inundated with the physical labor, financial obligations, and responsibilities their counterparts have struggled with. Health care and dental care are services needed by all. Medical students must labor after their university studies are completed or be required to pay back their debt.

Society makes some professions less important than others, and this is an error, for societies evolve through all contributors working together. If someone is not doing any form of work, then they should be reprimanded by the whole society. Non-laboring people must become contributors to our society's advancements by joining the workforce because we're inviting their participation.

It's through participating that one learns and begins to feel important in the cause. Society needs to stop giving handouts to people who are not maimed, blind, or handicapped because we're perpetuating the problem of non-laboring homeless people using illegal drugs abusing themselves.

All our oceans need to be cleaned up of all the waste. Billions of tons of plastic in the Pacific Ocean need to be funneled into Baja, California, to be loaded onto a conveyor train moving waste to a massive recycling complex. The recycling complexes will operate along the border between the United States and Mexico. We can always transform a fence into a train platform in a few years. The recycling complexes located in each state along the border will work as one big processing facility. Interconnected recycling complexes in the middle of the desert will process billions of tons of trash annually. The area between the mountain ranges in the center will be the prisoner complex where recycling takes place. America and Mexico will set aside their differences to clean up the world's environment, thereby protecting all our oceans' species. The US government and the Mexican government will share the recycling facilities' profits generated by hard prisoner labor. Interpol can also have prisoners sent to the recycling facilities to labor so international communities can contribute to the recycling project.

Prisoners will separate trash twenty-four hours a day, seven days a week, for manufacturing facilities needing raw materials. The raw materials collected will be transported to the companies needing them. The systems used at port facilities to scan freight looking for any temperature variations will be used to deter prisoner escape attempts. Small drones, solar gun towers, and robotic sensors, along with razor wire and other security designs, will line the mountains' perimeters to prevent escapes.

Some recycled materials can be sent to facilities manufacturing items for the Mars colonization. I'm initiating the "global zero-waste policy" so that everything must be recycled in every form. Every nation is held accountable for the stewardship of this planet. We all benefit from Earth sustaining us, so we're all responsible for its maintenance. All Earth's biodiversity, from its zoological species to its marine organisms, is our responsibility as human beings must preserve and protect our planet.

Each prisoner will be paid according to the weight of trash they separate. The harder a prisoner works, the higher the stipend. Prisoner stipends can be used in the prison commissary, on education expenses, and for personal savings. Prisoners who have a chance of parole will have saved assets upon their release, lessening the financial burden on society.

Prisoners serving life sentences will have permanent jobs in recycling as part of their punishment. Prisoners will be limited in the amount of funds family members or friends are able to give them to halt prison corruption and prisoner manipulation. Prisoners bribing guards will be non existent because of the selection process. The prisoners of various incarceration levels can be separated to perform different forms of labor, depending on the type of work required.

Prisoners up for parole can be trusted with more difficult recycling jobs than those who aren't eligible for parole at all. Trusted prisoners may process metals, glass, electronics, and automobiles, while prisoners who aren't trusted may process plastics, paper, wood, and compost. The point is that Texas, New Mexico, Arizona, California, and Mexico can build one big high-speed train for recycling and stop the pollution entering our waterways. All trash from America's fifty states, Mexico, and foreign countries can be processed efficiently in the recycling facilities. The recycling facilities will help keep all border areas free of debris, for we must begin beautifying America as well as Mexico. Other states can pay to dispose of their trash in the Southern states rather than building landfills, which increase toxicity within cities.

Mexicans can also be employed in recycling while they wait for their citizenship papers to process before they can enter the United States legally. Released prisoners can find employment in sanitation, operating equipment for waste recycling companies. Prisoners can find work in park establishment, fertilizer compost processing, agriculture, disaster cleanup, and construction. Don't be an outsider but be a committed insider for progress. We have to be careful that the liberals aren't being paid under the table by criminal organizations to help facilitate a prevention of punishment for those incarcerated for their crimes. Prisoners incarcerated need labor to provide a productive service back to the communities they've harmed and to reduce their destructive behaviors. Prison work crews are not an inhumane form of punishment but quite the contrary in prison populations where idle hands can lead to increased violence amongst prisoners. Prisoners will work their daily shifts in recycling then be able to return to their cells for dining and cell block recreation activities.

EPILOGUE

When disasters happen, celebrities and entrepreneurs come together, sending aid to areas of need. No one should be judgmental of celebrity portfolios. Celebrities often support hospital projects, surgeries, disaster relief, food banks, charity fundraisers, universities, inspirational films, housing development, political television, educational documentaries, and charity music events.

This book is intended to protect you from those who strive to hurt peaceful people like yourselves by giving you new ways for exploration. My Lord Jesus Christ has always been my helper and has opposed all those who oppose me. If your life is based on not harming anyone, that's a great life. Quickly discover those around yourself who have mental illness shielding your mind from their invasions of privacy. Your enemies will steal your written information and artistically engineered or architected possessions without restraint. When someone hates you there's nothing to deter them from wanting to ham your soul, talent, and spiritual gifts repetitively. Beware of those reciting biblical scripture in an attempt to sway your mind to be in their favor. You'll discover the two faced deceivers by investigating if they're living like Jesus Christ called believers to live as in the biblical scriptures. Often the most sinful people are those judging everyone else concealing inwardly sin into the deepest depths of their hearts. There are many deceivers in this word living two completely opposite lifestyles masquerading around like they're saints. It took me half my life to discover that each and everyone one of my siblings concealed Satan in the depth of their hearts working in union to war against my soul's spirit. Don't become a victim like me, for a true saint will live the Holy biblical scriptures in their words, deeds, thoughts, and works. Remember that the Devil's servants want to rip away all wealth, prosperity, talent, knowledge, power, love, and abilities you have to serve their wicked desires. There are many demonic people challenging themselves with goals which seek to achieve how many people they can deceive by living their lies very precisely while using Holy Bible as camouflage to conceal their many sins. Why are family members asking about where you keep your things? Why are family members prying into your life unless they've created sinful greed in the depths of their shallow hearts seeking to rob you of your hard worked on blessings that Yahweh has given you. It isn't hard figuring out a demon's plan especially when their every thought resides outside of obeying holy biblical scripture. The rings of power in the kingdom of heaven can only be wielded by those who know how evilly wicked demons are. Demons hide what they steal and they think this gives them power over the souls Yahweh has created. Demons do not turn aside from their deceitful ways they only try so much harder to convince others with more lies hidden in their masterful ways of deception. My advice is don't be deceived by those in worldly authority positions having uniforms or rank. Throughout world history many have betrayed Jesus Christ betraying souls with all their undermining thoughts. Jesus Christ knew the thoughts of those opposing him in the same way you will know the thoughts of others wanting you to have nothing. Many there will be who think they're entitled to a heavenly eternal life but all the things they've done in concealment are written in heaven and they cannot get away with what they want. For the Angels and Saints have the authority of Yahweh to wield power over the inhabitants of this world but many demonic people refuse to accept this judgment.

If you take only one thing from this book, I hope you visit Israel after you've filled yourself with all the scriptures of the Torah including the New Testament of the Holy Bible. The Holy Land has many hidden blessings that can be found when you know Holy Scripture. You must believe in Jesus Christ and in all the eternal words of the Holy Bible.

Please place all your trust in Yahweh and be one hundred (100) percent respectful while visiting the Holy Land. Yahweh is a jealous God; don't admire others in this world, for they can lead you into sinful paths. Biblical Scripture can help guide you to be respectful toward religious sanctuaries and all religions in Israel. When people from many faiths kneel acknowledging the supreme Creator of the universe they're on the right path in their lives and we shouldn't disrupt or deter them from seeking the truth. Many priests, nuns, and lay members of the church can guide you on a Holy Land pilgrimage visit.

Should you decide to visit Bethlehem in Palestine, my only advice is to be cautious of those who may behave like extortioners in an aggressive manner. I've experienced this extortionist behavior as a male visiting Bethlehem. So if you're a woman desiring to travel to Bethlehem, do it in a church-organized van or bus group for your safety. Avoid all confrontations in the Holy Land as well as when visiting holy sites. Be peaceful like Saint Joseph protecting the Virgin Mary with baby Jesus our Lord. Try to learn the languages spoken in Israel, which is very important. Don't limit yourself to learning only the Arabic languages; learn Hebrew as well as Latin to understand the holy masses.

The Holy Bible teaches us to not be offensive toward others but to sow kindness without strife. I have done what scripture says to do bring the complaints first to the offenders, then to the church leaders, and finally to the altar. We do not revolve around what demons want or think for they're the ones committing sin against us defying Yahweh's Holy commandments spoken for humanity to be obedient of for eternity. Beware of siblings with a belief that their sinful minds have total dominion over every tangent of your spirit. Yahweh made all people different. For demonic minds to think they're above you for all eternity this is the way of the Devil challenging Yahweh's authority to give you eternal life. Be willing to share with the needy throughout the Holy Land when you're able to do so. When in the Holy Land, behave as if Jesus was in your presence by being fruitful in all your conduct. Try to memorize the beatitudes from the Holy Bible living them.

I also suggest you take as few valuables as possible when visiting the Holy Land. Valuables only become a burden to care for during your stay. Your focus shouldn't be on your valuables but on the Lord Jesus Christ and Yahweh during your visit to the Holy Land.

Things that are common in the United States aren't so common in Israel, so pack the types of hygiene items you'll typically use for the duration of your trip. I hope your visit to the Holy Land will bring you closer to our Lord Jesus Christ. You're being summoned to walk on the water with Jesus, so take up your crosses and become fruitful spirits. I hope you go out into the world to create your eternal legacy making the biggest positive impact you can through hard work.

Peace be with you on your journey to a heavenly eternal life. I thank you for giving me your time, and I hope your spiritual journey will be forever blessed!

SINCERELY YOURS,
David P. Sarmiento

CPSIA information can be obtained
at www.ICGtesting.com
Printed in the USA
BVHW011551070721
611221BV00038B/617